Belonging

What does it mean to call a place home? How do we create community? When can we say that we truly belong?

Heartfelt issues of connection and community are the subject of bell hooks' new book, *Belonging: A Culture of Place*. Moving from past to present, hooks charts a journey in which she moves from place to place, from country to city and back again, only to end where she began in her native place — Kentucky.

In this provocative book, hooks explores a geography of the heart. She focuses on issues of homeplace, of land and land stewardship, linking these issues to local and global environmentalism and sustainability. Naturally, it would be impossible to contemplate these issues without the politics of race, gender, and class. hooks writes about ecology, sustainability, and finding solace in nature. She writes about family and the ties that bind. She focuses on the experiences of black farmers, past and present, who celebrate local organic food production.

With boldness, insight, and honesty, *Belonging* offers a remarkable vision of a world where all people — wherever they may call home — can live fully and well, where everyone can belong.

About bell hooks

bell hooks is a writer and cultural critic. Among her many books are the feminist classic *Ain't I A Woman*, the dialogue (with Cornel West) *Breaking Bread*, the children's books *Happy to Be Nappy* and *Be Boy Buzz*, the memoir *Bone Black* (Holt), and the general interest titles *All About Love*, *Rock My Soul*, and *Communion*. She has published six books with Routledge: *We Real Cool*, *Where We Stand*, *Outlaw Culture*, *Reel to Real*, *Teaching to Transgress*, and *Teaching Community*. Readers can look forward to her latest book in the teaching trilogy *Teaching Critical Thinking: Engaged Pedagogy*. Currently, she is Distinguished Professor in Residence in Appalachian Studies at Berea College in Kentucky.

Belonging:
A CULTURE OF PLACE

bell hooks

Routledge
Taylor & Francis Group

NEW YORK AND LONDON

First published 2009
by Routledge
711 Third Avenue, New York, NY 10017, USA

Simultaneously published in the UK
by Routledge
2 Park Square, Milton Park, Abingdon, Oxon, OX14 4RN

Routledge is an imprint of the Taylor & Francis Group, an informa business

© 2019 Taylor & Francis

Typeset in Bembo, Franklin Gothic, and Clarendon by codeMantra

Library of Congress Cataloging-in-Publication Data
hooks, bell.
Belonging: a culture of place / bell hooks.
p. cm.
Includes bibliographical references and index.
1. Home—Social aspects. 2. Home—Kentucky. 3. hooks, bell. 4. African American women—Kentucky—Biography. 5. African Americans—Kentucky—Biography.
6. Kentucky—Biography.
I. Title.
HQ503.H76 2008
305.48'8960730769092—dc22
[B]
2008021846

ISBN10: 0-415-96815-1 (hbk)
ISBN10: 0-415-96816-X (pbk)
ISBN10: 0-203-88801-4 (ebk)

ISBN13: 978-0-415-96815-7 (hbk)
ISBN13: 978-0-415-96816-4 (pbk)
ISBN13: 978-0-203-88801-8 (ebk)

To dancing in a circle of love — to living in beloved community

I am grateful for everyone in Berea for welcoming me — for giving me a place to belong —

And I especially give thanks for Pete Carpenter, Paige Cordial, Timi Reedy, Jane Post, Vicky and Clarence Hayes, Bobby Craig, Eugene Powell, Susan King, Stephanie Browner, Linda, Alina, Libby, Peggy, Tammy, Mayor Steve Connelly, Vernon, Angela and all my Kentucky family.

contents

acknowledgments

Chapter 4, "Touching the Earth" appeared first in *Sisters of the Yam: Black Women and Self-Recovery* (South End Press, 1993) © 1993 by Gloria Watkins.

Chapter 8, "Representations of Whiteness in the Black Imagination" appeared first in *Black Looks: Race and Representation* (South End Press, 1992) © 1992 by Gloria Watkins.

Chapter 10, "Earthbound: On Solid Ground" appeared first in *The Colors of Nature: Culture, Identity, and the Natural World*, edited by Alison Hawthorne Deming and Lauret E. Savoy (Milkweed Editions, 2002).

Chapter 11, "An Aesthetic of Blackness: Strange and Oppositional" and Chapter 14, "Aesthetic Inheritances: History Worked by Hand" appeared first in *Yearning: Race, Gender, and Cultural Politics* (South End Press, 1990) © 1990 by Gloria Watkins.

Chapter 12, "Inspired Eccentricity" by bell hooks, from *Family: American Writers Remember Their Own*, by Sharon Sloan Fiffer and Steve Fiffer, copyright © 1996 Sharon Sloan Fiffer and Steve Fiffer. Used by permission of Pantheon Books, a division of Random House, Inc.

Chapter 13, "A Place Where the Soul Can Rest" is reprinted by permission from *Etiquette: Reflections on Contemporary Comportment* edited by Ron Scapp and Brian Seitz, the State University of New York Press © 2007, State University of New York. All rights reserved.

1

Preface: To Know Where I'm Going

The idea of place, where we belong, is a constant subject for many of us. We want to know whether it is possible to live on the earth peacefully. Is it possible to sustain life? Can we embrace an ethos of sustainability that is not solely about the appropriate care of the world's resources, but is also about the creation of meaning — the making of lives that we feel are worth living. Tracy Chapman sings lyrics that give expression to this yearning, repeating, "I wanna wake up and know where I'm going." Again and again as I travel around I am stunned by how many citizens in our nation feel lost, feel bereft of a sense of direction, feel as though they cannot see where our journeys lead, that they cannot know where they are going. Many folks feel no sense of place. What they know, what they have, is a sense of crisis, of impending doom. Even the old, the elders, who have lived from decade to decade and beyond, say life is different in this time, "way strange," that our world today is a world of "too much" — that this too muchness creates a wilderness of spirit, the everyday anguish that shapes the habits of being for those who are lost, wandering, searching.

Mama's mama Baba (Sarah Oldham) would say a world of "too much wanting and too much waste." She lived a simple life, a life governed by seasons, spring for hoping and planting, summer for watching things grow, for walking and sitting on the porch, autumn for harvest and gathering, deep winter for stillness, a time for sewing and rest. All my childhood and into my first year of being grown up and living away from family, Baba lived secure in the two-story wood frame house that was her sanctuary on this earth, her home-place. She did not drive. No need to drive if you want your place on earth to be a world you can encompass walking. There were other folks like her in the world of my growing up, folks who preferred their feet walking solidly on the earth to being behind the wheel of an automobile. In childhood we were fascinated by the walkers, by the swinging arms and wide strides they made to swiftly move forward, covering miles in a day but always walking a known terrain, leaving, always coming back to the known reality, walking with one clear intent — the will to remain rooted to familiar ground and the certainty of knowing one's place.

Like many of my contemporaries I have yearned to find my place in this world, to have a sense of homecoming, a sense of being wedded to a place. Searching for a place to belong I make a list of what I will need to create firm ground. At the top of the list I write: "I need to live where I can walk. I need to be able to walk to work, to the store, to a place where I can sit and drink tea and fellowship. Walking, I will establish my presence, as one who is claiming the earth, creating a sense of belonging, a culture of place." I also made a list of places where I might like to dwell: Seattle, San Francisco, Tucson, Charleston, Santa Fe (these were just a few of the places on my list). I travel to them in search of that feeling of belonging, that sense that I could make home here. Ironically, my home state of Kentucky was not on the list. And at the time it would never have occurred to me,

not even remotely, to consider returning to my native place. Yet ultimately Kentucky is where my journey in search of place ends. And
where these essays about place began.

Belonging: A Culture of Place chronicles my thinking about issues
of place and belonging. Merging past and present, it charts a repetitive circular journey, one wherein I move around and around, from
place to place, then end at the location I started from — my old
Kentucky home. I find repetition scary. It seems to suggest a static
stuck quality. It reminds me of the slow languid hot summer days
of childhood where the same patterns of life repeat over and over.
There is much repetition in this work. It spans all my life. And it
reminds me of how my elders tell me the same stories over and over
again. Hearing the same story makes it impossible to forget. And so
I tell my story here again and again and again. Facts, ideas repeat
themselves as each essay was written as a separate piece — a distinct
moment in time.

Many of these essays in this book focus on issues of land and
land ownership. Reflecting on the fact that ninety percent of all
black folk lived in the agrarian South before mass migration to
northern cities in the early nineteen-hundreds, I write about black
farmers, about black folks who have been committed both in the
past and in the present to local food production, to growing organic
and to finding solace in nature. Naturally it would be impossible
to contemplate these issues without thinking of the politics of race
and class. It would be impossible to write about Kentucky's past
without bringing into the light the shadowy history of slavery in
this state and the extent to which the politics of racial domination
informs the lives of black Kentuckians in the present. Reflecting
on the racism that continues to find expression in the world of
real estate, I write about segregation in housing, about economic
racialized zoning. And while these essays begin with Kentucky as
the backdrop, they extend to politics of race and class in our nation
as a whole.

Similarly the essays focusing on the environment, on issues of sustainability reach far beyond Kentucky. Highlighting ways the struggle to restore balance to the planet by changing our relationship to nature and to natural resources. I explore the connections between black self-recovery and ecology. Addressing the issue of mountain-top removal, I write about the need to create a social ethical context wherein the concerns of Appalachians are deemed central to all American citizens. I write here about family, creating a textual album where I recall the folk who raised me, who nurtured my spirit.

Coming home, I contemplate issues of regionalism exploring my understanding of what it means to be a Kentucky writer. This collection of essays finds completion in my conversation with the visionary Kentucky writer, poet, essayist, and cultural critic Wendell Berry. Away from Kentucky I discovered Wendell's writings during my first year in college. What excited me most about him was his definitive commitment to poetry (at that time poetry was the central focus of my own writing). Yet he explored a wide range of issues in his essays that were fundamentally radical and eclectic. Following in Wendell's footsteps was from the start a path that would lead me back to my native place, to Kentucky. The first class I taught at Berea College focused on Berry's discussion of the politics of race in *The Hidden Wound*. In our conversation we reflect on this work, on his life and my own, the ways our paths converge despite differences of age and race.

On the journey to Wendell's farm in Port Royal, Kentucky, I saw many beautiful barns storing recently harvested tobacco. These images were the catalyst for the short reflection on the tobacco plant included in this collection.

Naming traits that he sees as central to Kentucky in his work *Appalachian Values*, Loyal Jones emphasizes the importance of family, commenting: "We think in terms of persons, we remember the people with whom we are familiar, and we have less interest in abstractions and people we have only heard about." Certainly many of the essays in *Belonging* begin with the family and kin with whom I am most familiar, especially in the essays focusing on creativity, aesthetic, and imaginative process. Writing about the past often places one at risk for evoking a nostalgia that simply looks back with longing and idealizes. Locating a space of genuineness, of integrity, as I recall the past and endeavor to connect it to the ideals and yearnings of the present has been crucial to my process. Using the past as raw material compelling me to think critically about my native place, about ecology and issues of sustainability, I return again and again to memories of family. During the writing of these essays, Rosa Bell, my mother, began to lose memory, to move swiftly into a place of forgetfulness from which there is no return. Witnessing her profound and ongoing grief about this loss, I learn again and again how precious it is to have memory.

We are born and have our being in a place of memory. We chart our lives by everything we remember from the mundane moment to the majestic. We know ourselves through the art and act of remembering. Memories offer us a world where there is no death, where we are sustained by rituals of regard and recollection. In *Belonging: A Culture of Place* I pay tribute to the past as a resource that can serve as a foundation for us to revision and renew our commitment to the present, to making a world where all people can live fully and well; where everyone can belong.

2

Kentucky Is My Fate

If one has chosen to live mindfully, then choosing a place to die is as vital as choosing where and how to live. Choosing to return to the land and landscape of my childhood, the world of my Kentucky upbringing, I am comforted by the knowledge that I could die here. This is the way I imagine "the end": I close my eyes and see hands holding a Chinese red lacquer bowl, walking to the top of the Kentucky hill I call my own, scattering my remains as though they are seeds and not ash, a burnt offering on solid ground vulnerable to the wind and rain — all that is left of my body gone, my being shifted, passed away, moving forward on and into eternity. I imagine this farewell scene and it solaces me; Kentucky hills were where my life began. They represent the place of promise and possibility and the location of all my terrors, the monsters that follow me and haunt my dreams. Freely roaming Kentucky hills in childhood, running from snakes and all forbidden outside terrors both real and imaginary, I learn to be safe in the knowledge that facing what I fear and moving beyond it will keep me secure. With this knowledge I nurtured a sublime trust in the power of nature to seduce, excite, delight, and solace.

Nature was truly a sanctuary, a place of refuge, a place for healing wounds. Heeding the call to be one with nature, I returned to the one state where I had known a culture of belonging. My life in Kentucky, my girlhood life, is divided into neat lines demarcating before and after. Before is the isolated life we lived as a family in the Kentucky hills, a life where the demarcations of race, class, and gender did not matter. What mattered was the line separating country and city — nature mattered. My life in nature was the Before and the After was life in the city where money and status determined everything. In the country our class had no importance. In our home we were surrounded by hills. Only the front windows of our house looked out on a solitary road constructed for the men seeking to find oil; all other windows faced hills. In our childhood, the rarely traveled road held no interest. The hills in the back of our house were the place of magic and possibility, a lush green frontier, where nothing manmade could run us down, where we could freely seek adventure.

When we left the hills to settle in town where the schools were supposedly better, where we could attend the big important church, Virginia Street Baptist (all things we were told would make us better, would make it possible for us to be somebody), I experienced my first devastating loss, my first deep grief. I wanted to stay in the solitude of those hills. I longed for freedom. That longing was imprinted on my consciousness in the hills that seemed to declare that all sweetness in life would come when we seek freedom. Folks living in the Kentucky hills prized independence and self-reliance above all traits.

While my early sense of identity was shaped by the anarchic life of the hills, I did not identify with being Kentuckian. Racial separatism, white exploitation and oppression of black folks, was so widespread it pained my already hurting heart. Nature was the place where one could escape the world of manmade constructions of race and identity. Living isolated in the hills we had very little contact with the world of white dominator culture. Away from the hills dominator culture and its power over our lives were constant.

Back then all black people knew that the white supremacist State with all its power did not care for the welfare of black folks. What we had learned in the hills was how to care for ourselves by growing crops, raising animals, living deep in the earth. What we had learned in the hills was how to be self-reliant.

Nature was the foundation of our counterhegemonic black sub-culture. Nature was the place of victory. In the natural environment, everything had its place including humans. In that environment everything was likely to be shaped by the reality of mystery. There dominator culture (the system of imperialist white supremacist capitalist patriarchy) could not wield absolute power. For in that world nature was more powerful. Nothing and no one could completely control nature. In childhood I experienced a connection between an unspoiled natural world and the human desire for freedom.

Folks who lived in the hills were committed to living free. Hillbilly folk chose to live above the law, believing in the right of each individual to determine the manner in which they would live their lives. Living among Kentucky mountain folk was my first experience of a culture based on anarchy. Folks living in the hills believed that freedom meant self-determination. One might live with less, live in a makeshift shack and yet feel empowered because the habits of being informing daily life were made according to one's own values and beliefs. In the hills individuals felt they had governance over their lives. They made their own rules.

Away from the country, in the city, rules were made by unknown others and were imposed and enforced. In the hills of my girlhood white and black folks often lived in a racially integrated environment, with boundaries determined more by chosen territory than race. The notion of "private property" was an alien one; the hills belonged to everyone or so it seemed to me in my childhood. In those hills there was nowhere I felt I could not roam, nowhere I could not go.

Living in the city I learned the depths of white subordination of black folks. While we were not placed on reservations, black folks were forced to live within boundaries in the city, ones that were not formally demarcated, but boundaries marked by white supremacist violence against black people if lines were crossed. Our segregated black neighborhoods were sectioned off, made separate. At times they abutted the homes of poor and destitute white folks. Neither of these groups lived near the real white power and privilege governing all our lives.

In public school in the city we were taught that Kentucky was a border state, a state that did not take an absolute position on the issue of white supremacy, slavery, and the continued domination of black folks by powerful whites. In school we were taught to believe that Kentucky was not like the deep South. No matter that segregation enforced by violence shaped these institutions of learning, that schools took children regularly to the Jefferson Davis monument, to places where the confederacy and the confederate flag was praised. To black folks it seemed strange that powerful Kentucky white folks could act as though a fierce white supremacy did not exist in "their" state. We saw little difference between the ways black folks were exploited and oppressed in Kentucky and the lives of black folks in other parts of the South, places like Alabama, Mississippi, and Georgia. By the time I graduated from high school my yearning to leave Kentucky had intensified. I wanted to leave the fierce racial apartheid that governed the lives of black folks. I wanted to find the place of freedom.

Yet it was my flight from Kentucky, my traveling all the way to the west coast, to California, that revealed to me the extent to which my sense and sensibility were deeply informed by the geography of place. The year I began my undergraduate education at Stanford University there were few students coming there from the state of Kentucky. I was certainly the only black student from Kentucky. And the prevailing social mores of racism meant that white Kentuckians did not seek my company. It was during this first year at Stanford that I realized the stereotypes about Kentucky that prevailed in the world beyond our region. Few folks there, at Stanford, knew anything about life in Kentucky. Usually, when asked where I heralded from, naming Kentucky as my home state would be greeted with laughter. Or with the question "Kentucky — where is that?"

Every now and then in those undergraduate years I would meet a fellow student who was sincere in their desire to hear about life in Kentucky, and I would talk about the natural world there, the lushness of our landscape, the waterfall at Blue Lake where I played as a child. I would talk about the caves and the trails left by the displaced Cherokee Indians. I would talk about an Appalachia that was black and white, about the shadow of coal dust on the bodies of black men coming home from working in the mines. I would talk about fields of tobacco, about the horses that make the Kentucky bluegrass a field of enchantment. I would talk with pride about the black male jockeys who were at the center of the horse racing events before imperialist white supremacist capitalist imposed rigid rules of racial segregation forcing black folks away from the public world of Kentucky horse culture.

Separating black folks, especially black jockeys, from the world of Kentucky horse culture went hand in hand with the rise in white supremacist thinking. For us it meant living with a culture of fear where we learned to fear the land, the animals, where we became fearful of the moist munching mouths of horses black jockeys would rarely ride again. This separation from nature and the concomitant fear it produced, fear of nature and fear of whiteness was the trauma shaping black life. In our psycho history, meaning the culture of southern black folk living during the age of fierce legally condoned racial apartheid, the face of terror will always be white. And symbols of that whiteness will always engender fear. The confederate flag, for example, will never stand for heritage for black folks. It still awakens fear in the minds and imaginations of elder black folks for whom it signaled the support of white racist assault on blackness.

White folks who mask their denial of white supremacy by mouthing slogans like "heritage not hate" to support their continued allegiance to this flag fail to see that their refusal to acknowledge what this "heritage" means for black folks is itself an expression of white racist power and privilege. For the confederate flag is a symbol of both heritage and hate. The history of the confederacy will always evoke the memory of white oppression of black folks with rebel flags, guns, fire, and the hanging noose — all symbols of hate. And even though many poor and disenfranchised white Kentuckians struggling to make their way through the minefield of capitalist white power mimic and claim this history of colonial power, they can never really possess the power and privilege of capitalist whiteness. They may embrace this symbol to connect them to that very world and that past that denied their humanity but it will never change the reality of their domination by those very same forces of white supremacist hegemony.

Growing up in that world of Kentucky culture where the racist aspect of the confederate past was glorified, a world that for the most part attempted to obscure and erase the history of black Kentuckians, I could not find a place for myself in this heritage. Even though I can now see in retrospect that there were always two competing cultures in Kentucky, the world of mainstream white supremacist capitalist power and the world of defiant anarchy that championed freedom for everyone. And the way in which that culture of anarchy had distinct anti-racist dimensions accounts for the unique culture of Appalachian black folks that is rarely acknowledged. It is this culture Loyal Jones writes about in *Appalachian Values* when he explains: "Many mountaineers, as far South as Alabama and Georgia were anti-slavery in sentiment and fought for the Union in the Civil War, and although Reconstruction legislatures imposed anti-Negro laws, thus training us in segregation, Appalachians for the most part, have not been saddled with the same prejudices against black people that other southerners have." Even though I spent my early childhood around mountain white folks who did not show overt racism, and that this world of racial integration in Kentucky hills had been a part of my upbringing, shaping my sense and sensibility, our move away from that culture into the mainstream world and its values meant that it was white supremacy that shaped and informed the nature of our lives once we were no longer living in the hills. It was this legacy of racial threat and hate that engendered in me the desire to leave Kentucky and not return.

Leaving Kentucky I believed I would leave the terror of whiteness behind but that fear followed me. Away from my native place I learned to recognize the myriad faces of racism, racial prejudice and hatred, the shape-shifting nature of white supremacy. During my first year at Stanford University, I felt for the first time the way in which geographical origins could separate citizens of the same nation. I did not feel a sense of belonging at Stanford University; I constantly felt like an unwanted outsider. Just as I found solace in nature in Kentucky, the natural environment, trees, grass, plants, the sky in Palo Alto, California all offered me a place of solace. Digging in the California ground my hands touched earth, that was so different from the moist red and brown dirt of Kentucky: I felt awe. Wonder permeated my senses as I pondered the fact that traveling thousands of miles away from my native place had actually changed the very ground under my feet. Then I could not understand how the earth could be my witness in this strange land if it could not be a mirror into which I could see reflected the world of my ancestors, the landscape of my dreams. How could this new land hold me upright, provide me the certainty that the ground of my being was sound?

Seeking an experience of intellectual life in the academic world I entered an environment based on principles of uncertainty, an opportunistic world where everything changes and ends. I longed to return to my native place where there was fierce reluctance to accept change. Kentucky is one of the states in our nation known for its hard-headed refusal to embrace change. In the old days most Kentucky folks wanted everything to pass down from generation to generation unchanged. This refusal to promote change was most evident in the arena of race relations. White folks in the state of

Kentucky kept racial segregation the norm long after other states had made significant progress in the direction of civil rights.

Conservative white Kentuckians told themselves "the blacks don't want change — they like the way things are." Having reduced black folks to a state of traumatic powerlessness, racist white folks saw no problem with the intimate racial terrorism that they enacted which led them to believe that they could know the mind and hearts of black folks, that they could own our desires. Efforts on the part of conservative white Kentuckians to exploit and oppress black folk were congruent with the effort to erase and destroy the rebellious sensibility of white mountain folk. The anarchist spirit that had surfaced in the culture of white hillbillies was as much a threat to the imperialist white supremacist capitalist state as any notion of racial equality and racial integration. Consequently this culture, like the distinctive habits of black agrarian folk, had to be disrupted and ultimately targeted for destruction.

Leaving Kentucky, fleeing the psycho history of traumatic powerlessness, I took with me from the sub-cultures of my native state (mountain folk, hillbillies, Appalachians) a positive understanding of what it means to know a culture of belonging, that cultural legacy handed down to me by my ancestors. In her book *Rebalancing the World* Carol Lee Flinders defines a culture of belonging as one in which there is "intimate connection with the land to which one belongs, empathic relationship to animals, self-restraint, custodial conservation, deliberateness, balance, expressiveness, generosity, egalitarianism, mutuality, affinity for alternative modes of knowing, playfulness, inclusiveness, nonviolent conflict resolution, and openness of spirit." All these ways of belonging were taught to me in my early childhood but these imprints were covered over by the received biased knowledge of dominator culture. Yet they become the subjugated knowledge that served to fuel my adult radicalism.

Living away from my native place I become more consciously Kentuckian than I was when I lived at home. This is what the experience of exile can do, change your mind, utterly transform one's perception of the world of home. The differences geographical location imprinted on my psyche and habits of being became more evident away from home. In Kentucky no one had thought I had a Kentucky accent; in California speaking in the soft black southern vernacular that was our everyday speech made me the subject of unwanted attention. In a short period of time I learned to change my way of speaking, to keep the sounds and cadences of Kentucky secret, an intimate voice to be heard only by folks who could understand. Not speaking in the tongue of my ancestors was a way to silence ridicule about Kentucky. It was a way to avoid being subjugated by the geographical hierarchies around me which deemed my native place country backwards, a place outside time. I learned more about Kentucky during my undergraduate years as I placed the portrait of a landscape I knew intimately alongside the stereotypical way of seeing that world as it was represented by outsiders.

Perhaps my greatest sense of estrangement in this new liberal college environment was caused by the overall absence among my professors and peers of any overtly expressed belief in Christianity and God. Indeed, it was far more cool in those days to announce that one was agnostic or atheist than to talk about belief in God. Coming from a Bible-toting, Bible-talking world where scripture was quoted in everyday conversations, I lacked the psychological resources and know-how to positively function in a world where spiritual faith was regarded with as much disdain as being from the geographical South. In my dormitory the one student who openly read from scripture, a quiet white male student from a Mormon background, was more often than not alone and isolated. We talked to one another and endeavored to make each other feel less like strangers in a strange land. We talked scripture. But talking scripture was not powerful enough to erase the barriers created by racism that had taught us to fear and beware difference. And even though there were organized Christian groups on campus they did not speak the religious language I was accustomed to hearing.

By the end of my second year of college, I began to question the religious beliefs of my family, the way of religion I had been taught back home. In the new age spiritual environment of California, I fashioned a spirituality that made sense to my mind and heart. I worshiped in a manner that was in tune with divine spirit as I had come to know it in the hills of my Kentucky upbringing. Growing up I had always been torn between the righteous religious fundamentalism of those who practiced according to organized church doctrine and dictates and the nature-worshiping ecstatic mystical spirituality of the backwoods. All through my college years, even during those times when my soul was racked by doubt, I held onto core beliefs in the power of divine spirit.

My college years began that process of feeling split in my mind and heart which characterized my life in all the places I moved to: California, Wisconsin, Connecticut, Ohio, New York. At heart I saw myself as a country girl, an eccentric product of the sense and sensibility of the Kentucky backwoods and yet the life I lived was one where different ethics, values, and beliefs ruled the day. My life away from Kentucky was full of contradictions. The issues of honesty and integrity that had made life clear and simple growing up were an uneasy fit with the academic and literary world I had chosen as my own. In time the split mind that had become my psychic landscape began to unravel. As I experienced greater success as an intellectual and a writer, I felt I was constantly working to make my core truths have visibility and meaning in a world where the values and beliefs I wanted to make the foundation of my life had no meaning. Still and all, I did not feel that I could come home. The self I had invented in these other worlds seemed too unconventional for Kentucky, too cosmopolitan.

Like many writers, especially southerners, who have stayed away from their native place, who live in a state of mental exile, the condition of feeling split was damaging, caused a breaking down of the spirit. Healing that spirit meant for me remembering myself, taking the bits and pieces of my life and putting them together again. In remembering my childhood and writing about my early life I was mapping the territory, discovering myself and finding homeplace — seeing clearly that Kentucky was my fate.

The intense suicidal melancholia that had ravished my spirit in girlhood, in part a response to leaving the hills, leaving a world of freedom, had not been left behind. It followed me to all the places I journeyed. And the familiar grief which kept me awake at night, crying, longing, stayed present wherever I went, bringing in its wake the experience of traumatic powerlessness. The nighttime terrors that were there in Kentucky, the wild horses that roamed my nights leaving me crazed and sleep deprived followed me. The inability to sleep that was a constant in girlhood was even more exacerbated the farther I journeyed from my childhood home. Many times I would lie in pitch dark rooms away from Kentucky and imagine all the ways I could create my own homeplace. Yet all my efforts to start over always ended up taking me back to the past, allowing it to serve as foundation for the present.

When in doubt about the direction of my life, I would imagine myself as a filmmaker, creating an autobiographical film titled *Kentucky Is My Fate*. The first frames of the film are all shots of nature, shots of tobacco fields, tobacco farms, tobacco barns. I enter this filmic narrative as a witness: Baba, Mama's mother, is braiding tobacco leaves, readying them to be hung, for placing in closets and trunks to keep moths and other cloth-eating pests away. Much of this imaginary film focuses on the elders whose presence dominated my childhood.

When I left my native place for the fist time, I brought with me two artifacts from home that were emblematic of my growing-up life, braided tobacco leaves and the crazy quilt Baba, Mama's mother, had given me when I was a young girl. These two totems were to remind me always of where I come from and who I am at my core. They stand between me and the madness that exile makes, the brokenheartedness. They are present in my new life to shield me from death, to remind me that I can always return home. Each year of my life as I went home to visit, it was a rite of passage to reassure myself that I still belonged, that I had not become so changed that I could not come home again. My visits home almost always left me torn: I wanted to stay but I needed to leave, to be endlessly running away from home.

Madness was more acceptable away from home. At the predominately white colleges I attended, it was accepted that students might feel overwhelmed by separation from their norm environment, that we might feel estranged, alienated, that we might in fact lose our minds. Therapy, I learned then, was the best way to face psychic wounds, the best way to heal. One of my younger sisters recently asked me: "How did you know you needed help?" I shared: "I knew I was not normal. I knew it was not normal to want to kill myself." Intensely sad suicidal longings led me to therapy but in those early years therapy did not help. I could not find a therapist who would acknowledge the power of geographical location, of ancestral imprints, of racialized identity. Watching the comedy *Beverly Hillbillies* seemed to be the basis for most folks' perceptions of the Kentucky backwoods, even therapeutic ones. And certainly in my early college years I lacked an adequate language to name all that had shaped and formed me.

Even when I felt therapy was not helping, I did not lose my conviction that there was health to be found, that healing could come from understanding the past and connecting it to the present. Baba, my maternal grandmother, would often ask me, "How can you live so far away from your people?" When she posed this question, I always felt it carried with it a rebuke, the slight insistence that I had been disloyal, betrayed the ancestral legacy by leaving home. The question I asked myself was, "Why when Kentucky means so much, why can't you go home and stay home?" In my early twenties I began to construct a narrative map of the past, to write down the experiences of childhood that I felt were vital imprints. I began by making list — thinking all the while about the stories we tell someone about ourselves when we meet and begin the process of getting to know one another. It was clear to me that I shared the same tales I thought were significant over and over again. I felt certain that if I could just put these memories on paper and order them it would help me to bring order to my life. Creating a clear detailed account of "myself," I felt certain that I would then be able to stand back and see myself in a new way, no longer fragmented — whole — complete.

Writing my girlhood life helped. It gave me new ground to stand on. I collected these memories and published them in the memoir *Bone Black*. Poetic in style and tone, abstract even, I read and hear these accounts of my girlhood as though the speaker is in a trance, in a state that is at once removed and yet present. Much of my life away from Kentucky was lived in a trance state, as though I was always there and not there at the same time. Working to heal, to be whole, has been a process of awakening, of moving from trance into reality, of learning how to be fully present. Leaving home evoked extreme feeling of abandonment and loss. It was like dying. Resurrecting the memories of home, bringing the bits and pieces together was a movement back that enabled me to move forward. All my trance states were defenses against the terrors of childhood. When I left home I took with me unresolved traumas. Carrying the voices of my ancestors within me everywhere I called home, I carried remembered pain and allowed it to continually sweep me away. This sensation of being swept away was like spinning.

Away from Kentucky my heart was spinning and it was only when the spinning stopped that I could see clearly and heal. Initially this clarity did not lead me to return to Kentucky. Indeed, I feared that if I returned home to Kentucky I would be shattered, triggered in ways that would disrupt and fragment. I could be most adamantly a Kentuckian away from Kentucky. Since my native place was indeed the site and origin of the deep dysfunction that had damaged my spirit I did not believe I could be safe there. I could see the connection between private family dysfunction, and the public dysfunction that was sanctioned by the State of Kentucky. Wayne Kritsberg offers a useful definition of dysfunction in his book *Healing Together* clarifying that: "A dysfunctional family is one that is consistently unable to provide a safe nurturing environment. Through its maladaptive behaviors, the family develops a set of restrictions that inhibit the social and emotional growth of its members, particularly the children. The healthy family on the other hand provides safety and nurturing for its members and assists them in their development by setting firm but reasonable limits, rather than imposing rigid constraints." The fundamentalist Christian patriarchal power that determined the public world of the State in my native place was mirrored in the structure of my primary family life and family values. Concurrently, white supremacy shaped the psyches of black and white folks in ways that constrained and deformed.

Making the connections between geographical location and psychological states of being was useful for me. It empowered me to recognize the serious dysfunctional aspect of the southern world I was raised in, the ways internalized racism affected our emotional intelligence, our emotional life, and yet it also revealed the positive aspects of my upbringing, the strategies of resistance that were life enhancing. Certainly racial separatism, in conjunction with resistance to racism and white supremacy, empowered non-conforming black folks to create a sub-culture based on oppositional values. Those oppositional values imprinted on my psyche early in childhood enabled me to develop a survivalist will to resist that stood me in good stead both during the times I returned home and in the wilderness of spirit I dwelled in away from home. Oppositional habits of being I had learned during childhood forged a tie to my native place that could not be severed

Growing up, renegade black and white folks who perceived the backwoods, the natural environment, to be a space away from manmade constructions, from dominator culture, were able to create unique habits of thinking and being that were in resistance to the status quo. This spirit of resistance had characterized much of Kentucky's early history, the way in which white colonizers first perceived it an untouched truly wild wilderness that would resist being tamed by the forces of imperialist white supremacist capitalism. Even though the forces of imperialist white supremacist capitalist patriarchy did ultimately subordinate the land to its predatory interests it did not create a closed system; individual Kentuckians white and black, still managed to create subculture, usually in hollows, hills, and mountains, governed by beliefs and values contrary to those of mainstream culture. The free-thinking and non-conformist behavior encouraged in the backwoods was a threat to imperialist white supremacist capitalist patriarchy, hence the need to undermine it by creating the notion that folks who inhabited these spaces were ignorant, stupid, inbred, ungovernable. By dehumanizing the hillbilly, the anarchist spirit which empowered poor folks to choose a lifestyle different from that of the state and so-called civilized society could be crushed. And if not totally crushed, at least made to appear criminal or suspect.

This spirit of resistance and revolution that has been nurtured in me by generations of Kentucky black folks who had chosen self-reliance and self-determination over dependency on any government provided the catalyst for my personal struggle for self-definition. The core of that resisting oppositional culture was an insistence on each of us being people of worth and dignity. Acknowledging one's worth meant that one had to choose to be a person of integrity, to stand by one's word. In my girlhood I was taught by my elders, many of whom had not been formally educated and lacked basic skills of reading and writing, that to be a person of integrity one had to always tell the truth and always assume responsibility for your actions. Particularly, my maternal grandmother Baba taught me that these values should ground my being no matter my chosen place or country. To live these values then, I would, she taught, need to learn courage — the courage of my convictions, the courage to own mistakes and make reparation, the courage to take a stand.

In retrospect I have often wondered whether her insistence on my always being dedicated to truth, a woman of my word, a woman of integrity, was the lesson learned by heart that would ultimately make it impossible for me to feel at home away from my native place, away from my people. Striving to live with integrity made it difficult for me to find joy in life away from the homefolk and landscape of my upbringing. And as the elders who had generously given of their stories, their wisdom, their lives to make it possible for me, and folks like me, to live well, more fully, began to pass away, it was only a matter of time before I would be called to remembrance, to carry their metaphysical legacy into the present. Among illiterate backwoods folk I had been taught values, given ethical standards by which to live my life. Those standards had little meaning in the world beyond the small Kentucky black communities I had known all my life.

If growing up in an extremely dysfunctional family of origin had made me "crazy," surviving and making home away from my native place allowed me to draw on the positive skills I had learned during my growing-up years. Kentucky was the only place I had lived where there were living elders teaching values, accepting eccentricity, letting me know by their example that to be fully self-actualized was the only way to truly heal. They revealed to me that the treasures I was seeking were already mine. All my longing to belong, to find a culture of place, all the searching I did from city to city, looking for that community of like-minded souls, was waiting for me in Kentucky, waiting for me to remember and reclaim. Away from my home state I often found myself among people who saw me as clinging to old-fashioned values, who pitied me because I did not know how to be opportunistic or play the games that would help me get ahead.

I am reminded of this tension causing duality of desire when I read Lorraine Hansberry's play *A Raisin in the Sun*. In the play she dramatizes the conflicts that emerge when the values of belonging, the old ways, collide with the values of enterprise, and career opportunism. Sad that her son wants to take the insurance money they have received at the death of her husband, Mama declares: "Since when did not money become life." Walter Lee answers: "It was always life mama. We just did not know it." No doubt masses of black folk fleeing the agrarian South for the freedom from racist exploitation and oppression they imagined would not be their lot in the industrialized North felt an ongoing conflict of values. Leaving the agrarian past meant leaving cultures of belonging and community wherein resources were shared for a culture of liberal individualism. There is very little published work that looks at the psychological turmoil black folks faced as they made serious geographical changes that brought with them new psychological demands.

Certainly when I left Kentucky with its old-fashioned values about how to relate in the world, I was overwhelmed by the lack of integrity I encountered in the world away from home. Most folks scoffed at the notion that it was important to be honest, to be a person of one's word. This lack of integrity seemed to surface all the more intensely when I moved to New York City to further my career as a writer. During these years away from my native place, I often felt confusion and despair. My fundamentalist Christian upbringing had taught me to consider the meaning of sin as missing the mark. During those times in my life I often felt I was missing the mark, failing to live in accordance with the core values I believed should be the foundation of my identity. I struggled psychologically to repair the damages to my soul inflicted by my trespasses and those who trespassed against me.

Becoming successful as a cultural critic and creative writer, away from my native place, I was consistently astounded when readers and reviewers who wrote about my work failed to mention the extent to which the culture of place I had known in Kentucky shaped my writing and my vision. Surprised when the literary world did not acknowledge the significance of my Kentucky roots, I felt a greater necessity to articulate the role of homeplace in my artistic vision. Often critics would talk about my southern roots, never naming a specific location for those roots. To some extent this failure to focus on Kentucky was linked to assumptions about whether Kentucky really was the "South." I would tell people that growing up black in Kentucky we experienced our world as southern, as not very different from other southern places, like Alabama and Georgia. It may very well be that the culture of whiteness in Kentucky has characteristics that would not have been seen as distinctly southern but certainly the sub-cultures black folks created and create were formed by the understanding of what it meant to be black people in the South. For all the talk about Kentucky as a border state, the culture of slavery, of racial apartheid, had won the day in the state despite places in the region that had sprouted fierce assertions of civil rights for all. Certainly, reading the biographical and autobiographical memoirs of black Kentuckians one learns of a world shaped by feudal forces of imperialist white supremacist capitalism but one also learns of all the inventive ways black folks deployed to survive and thrive in the midst of exploitation and oppression.

During the more than thirty years that I did not make my home in Kentucky, much that I did not like about life in my home state (the cruel racist exploitation and oppression that continued from slavery into the present day, the disenfranchisement of poor and/or hillbilly people, the relentless assault on nature) was swiftly becoming the norm everywhere. Throughout our nation the dehumanization of poor people, the destruction of nature for capitalist development, the disenfranchisement of people of color, especially African-Americans, the resurgence of white supremacy and with plantation culture has become an accepted way of life. Yet returning to my home state all the years that I was living away, I found there essential remnants of a culture of belonging, a sense of the meaning and vitality of geographical place.

All the positive aspects of a culture of belonging that Kentucky offered me were not present in other places. And maybe it would have been harder for me to return to my native place if I had not

consistently sustained and nurtured bonds of kin and family despite
living away. My last lengthy place of residence prior to becoming
a resident of Kentucky was New York City. Had anyone ever pre-
dicted when I was younger that I would one day live in Manhattan,
I would have responded, "That is never gonna happen — 'cause I am
a country girl through and through." Concurrently, had I been told
that I would return in mid-life to live in Kentucky, I would have
responded: "when they send my ashes home." New York City was
one of the few places in the world where I experienced loneliness for
the first time. I attributed this to the fact that there one lives in close
proximity to so many people engaging in a kind of pseudo intimacy
but rarely genuinely making community. To live in close contact
with neighbors, to see them every day but never to engage in fellowship was downright depressing. People I knew in the city often ridiculed the idea that one would want to live in community — what
they loved about the city was the intense anonymity, not knowing
and not being accountable. At times I did feel a sense of community
in the city and endeavored to live in the West Village as though it
were a small town. Bringing my Kentucky ways with me wherever I
made homeplace sustained my ties to home and also made it possible
for me to return home.

My decision to make my home in Kentucky did not emerge from
any sentimental assumption that I would find an uncorrupted world
in my native place. Rather I knew I would find there living remnants
of all that was wonderful in the world of my growing up. During
my time away, I would return to Kentucky and feel again a sense of
belonging that I never felt elsewhere, experiencing unbroken ties to
the land, to homefolk, to our vernacular speech. Even though I had
lived for so many years away from my people, I was fortunate that
there was a place and homefolk for me to return to, that I was welcomed. Coming back to my native place I embrace with true love
the reality that "Kentucky is my fate" — my sublime home.

3

Moved by Mountains

Life is full of peaks and valleys, triumphs, and tribulations. We often cause ourselves suffering by wanting only to live in a world of valleys, a world without struggle and difficulty, a world that is flat, plain, consistent. We resist the truth of difference and diversity. We resist acknowledging that our constants exist within a framework where everything is always changing. We resist change. When we are able to face the reality of highs and lows embracing both as necessary for our full development and self-actualization, we can feel that interior well-being that is the foundation of inner peace. That life of appreciation for difference, for diversity, a life wherein one embraces suffering as central to the experience of joy is mirrored for us in our natural environment.

Earth is a diverse ecosystem. Mountains, hills, valleys, rivers and lakes, the forest are all naturally organically balanced. We have much to learn as inhabitants, as witnesses to this environment. As the indigenous Native Americans who peopled the Americas before the rest of us believed, if we listen, nature will teach us. However, if we think of the natural landscapes that surround us as simply, blank slates, existing for humans to act upon them according to our will then we cannot exist in life-sustaining harmony with the earth. We cannot proudly declare like the biblical psalmist that "I will lift up mine eyes unto the hills from whence cometh my help." The psalmist wanted us to know that we can gain spiritual strength by simply beholding the natural world, that indeed to look upon the wonders of nature is to gaze at divine spirit. Estrangement from our natural environment is the cultural contest wherein violence against the earth is accepted and normalized. If we do not see earth as a guide to divine spirit, then we cannot see that the human spirit is violated, diminished when humans violate and destroy the natural environment.

Nothing epitomizes this violence more in our contemporary life than mountain-top removal (when the summit of the mountain is removed to extract coal) and the devastation that occurs in its wake. In Stephen George's essay "Bringing Down The Mountain," he explains the way it all happens: "Mountaintop-removal mining is a simple process, plow the trees (but don't bother to harvest them) and everything else living on the mountain, blast off the top (usually 800 to 1,000 feet), take out the coal, and leave a leveled area... ." Much of this mining takes place in Appalachia yet it still one of the materially poorest regions in our nature. The wealth that is in our natural world when measured in dollars is not ever abundant yet it could be so if humans were not abusing and wasting this precious resource. As George explains, the amazing natural legacy of the Appalachians is endangered: "a splendorous spread of rolling hills and green mountains mirrored nowhere in the world — is being systematically destroyed so than an unsustainable way of life in our cities may continue."

Coal is one of earth's great gifts. As a child in Kentucky our family lived in an old Victorian-style house. Its modern heating system was not effective. To stay warm during the freezing cold months we burned coal in the small fireplaces that were a given in this old-style architecture. Watching the coal burn, feeling its heat, we were in our childhood filled with wonder. Coal was awesome. Colored the deepest shade of black, it was both beautiful and functional. Yet it did not come into our homes and into our lives without tremendous sacrifice and risk.

In the early evenings when the neighborhood men who mined coal came home from work with their bodies covered in ash, their hats with lights, carrying their lunch boxes, we would follow them, not understanding that they were beat, bone weary, not in the mood to play. There is no child raised in the culture of coal mining who does not come to understand the risks involved in harvesting coal. In the world of coal mining without big machinery, coal mining has a human face. Man is limited in his physical capacity. He can only extract so much. Machines can take and keep taking.

The smallest child can look upon a natural environment altered by conventional mining practice and see the difference between that process and mountaintop removal. Introducing the collection of essays in the book *Missing Mountains*, Silas House shares the way in which being raised in a cold mining family was for him a source of pride. He begins with the statement, "coal mining is a part of me," then recalls a long history of family members working in the mines. And while he speaks against mountaintop removal, he shares this vital understanding: "We are not against the coal industry. Coal was mined for decades without completely devastating the entire region. My family is a part of that coal-mining legacy. But mountaintop removal means that fewer and fewer people work in mining, because it is so heavily mechanized. If mountaintop removal is banned, there might actually be more mining jobs for the hard-working people of Kentucky. And beyond that the proper respect might finally be returned to the spirit of the land and its people." Without a sustainable vision of coal usage, without education for creating consciousness that would enable our nation to break with unhealthy dependency on coal, we cannot restore the dignity both to the earth and to this rich resource.

Mountaintop removal robs the earth of that dignity. It robs the folk who live in the cultural wasteland it creates of their self-esteem and divine glory. Witnessing up close the way this assault on the natural environment ravages the human spirit, the anguish it causes folk who must face daily the trauma of mountaintop removal, we who live away from this process are called to an empathy and solidarity that requires that we lend our resources, our spiritual strength, our progressive vision to challenge and change this suffering.

A beacon light to us all, elder Daymon Morgan embodies the unbridled spirit of a true Kentucky revolutionary. He acts as a conservationist, a steward of the land, and as one who is committed to the struggle to end mountaintop removal. Returning from World War II, Morgan bought land on Lower Bad Creek in Leslie County, Kentucky. Raising a family, growing herbs on his land, he had allowed the earth to teach him, to be his witness. His is special because he is in many ways representative of the ordinary citizen who is called to political action because of their love of the land and community. In recent times the Appalachian Studies program at Berea College makes certain that the faculty and staff, especially those who are new, take the Appalachian tour so that they may better understand our region, have an opportunity to meet this amazing man of integrity who stands for all that is right and wonderful in a democratic country. Taking the tour provided me an opportunity to meet Mr. Morgan, to be in his presence, to learn from his knowledge. Even before he opens his mouth, the strength and stillness of his being radiates glory. In Buddhist tradition the student learns that it is transformative just to stand in the presence of a great teacher.

Both by his presentation and in my short dialogue with Mr. Morgan, I saw in his visage and heard in his own words the extent to which fighting mountaintop removal wears on his spirit, wears him down, especially when that resistance must take the form of challenging relatives who would surrender the land, their legacy to big business. Before meeting Daymon Morgan, I had learned from his writing about the tens of thousands of years it takes for the organic matter of the forest to biodegrade and make rich. When this earth is attacked he mourns: "It's very disturbing to me to see the things that I love being destroyed. I got my medicine and my food from these mountains, and I still do. There's a place down here where I can lay down and drink out of this creek and I want to keep it that way because it's clear above. I feel like I'm being pushed into a corner." Just two years later hearing Morgan speak we hear the emotional toil resistance takes. Yet he tells with pride that there is joy in struggle, that he continues to struggle because of the debt he owes this Kentucky land. He honors the mutual relationship between him and the earth by working to protect and preserve the world around him. I ask him about protecting this legacy beyond the grave. No matter the steps he does not take to still be resisting beyond death, his presence is making a difference in the here and now.

Unlike other Appalachian tour groups who have visited at Morgan's home, we were not able to make it up the mountain in our bus. He came down the mountain to talk with us. We were watched by coal-mining hired hands sitting in vehicles. Subjected to a level of surveillance that bordered on harassment, their intent was to block us using roads that would enable us to see first hand the devastation. Their intent was to keep us from seeing the work of mountaintop removal. Concern for our safety was paramount to Mr. Morgan. Still we were able to witness and experience the threat he faces daily from those who could care less about the survival of our Kentucky land, culture, and the lives of folks who are mostly poor and working class. The lack of empathy for the lives that are devastated by mountaintop removal reminds us of the overall crisis in human values generated by dominator culture, by imperialist white supremacist capitalist patriarchy.

In dominator culture the will to power stands as a direct challenge to the cultural belief that humans survive soulfully because of a will to meaning. When the will to meaning is paramount, human life retains dignity. The capacity of humans to create community, to make connections, to love, is nurtured and sustained. For those us who believe in divine spirit, in higher powers, the issue of mountaintop removal and all practices wherein the earth is plundered and the environment wasted is as much a spiritual issue as it is a political issue. In order to justify dehumanizing coal-mining practices, the imperial capitalist world of big business has to make it appear that the plant and human life that is under attack has no value. It is not difficult to see the link between the engrained stereotypes about mountain folk (hillbillies), especially those who are poor, representations that suggest that these folk are depraved, ignorant, evil, licentious, and the prevailing belief that there is nothing worth honoring, worth preserving about their habits of being, their culture.

Mass media representations of poor folk in general convey to the public the notion that poor people are in dire straits because of the bad choices they have made. It pushes images that suggest that if the poor suffer from widespread addiction to sugar, alcohol or drugs it is because of innate weaknesses of character. When mass media offers representations of poor mountain folk, all the negative assumptions are intensified and the projections exaggerated. No wonder then that is usually easier for citizens concerned about environmental issues to identify with the hardships facing nature and the lives of the poor in underdeveloped countries than to identify with the exploitation of the environment, both the natural and cultural world of people here in our society, especially in Appalachia.

In Alice Walker's most recent book, *We Are The Ones We Have Been Waiting For*, she describes soul murder in dominator culture as "pain that undermines our every attempt to relieve ourselves of external and internalized ... domination ... the pain that murders our wish to be free." She concludes, "It is a pain that seems unrelenting. A pain that seems to have no stopping and no end. A pain that is ultimately, insidiously, turning a generous life loving people into a people who no longer feel empathy for the world. We are being consumed by our suffering." While Walker is talking about the fate of black folk, her words speak to the human condition in our culture, especially to the lives of exploited and oppressed people of all colors.

To truly create a social ethical context wherein masses of American citizens can empathize with the life experiences of Appalachians we must consistently challenge dehumanizing public representations of poverty and the poor. Restoring to our nation the understanding that people can be materially poor yet have abundant lives rich in engagement with nature, with local culture, with spiritual values, is essential to any progressive struggle to halt mountaintop removal. Seeing and understanding that abundance means not only that we must collectively as a nation change our thinking about poverty, it means we must see a value in life that is above and beyond profit motives. And that is a challenging task in a material cultural where individual citizens of all classes spend significant amounts of their daily life fantasizing about becoming wealthy by winning the lottery as well as spending much of their income on the purchase of lottery tickets. This situation would be cause for widespread despair were it not for the education for critical consciousness that is already leading many American citizens to revaluate their lives. Among all classes, decreased economic resources caused by job loss, low wages, high housing costs etc. are all circumstances that are serving as a catalyst for folks to re-think their lives. This rethinking often includes a return to spiritual values which acts to reconnect folk with nature. Walker tells us in her recent work that we have only spirit to guide us, that "spirit is our country because it is ultimately our only home."

One of the unintended benefits that have come with the widespread rebirth of religious fundamentalism has been the outgrowth of new ways of thinking about the poor. Concurrently, this revived theology calls for those who are truly living according to the biblical word to identify with the poor and to seek to live simply. That call to simple living often begins with a reawakening of wonder sparking awareness of our profound connection to nature. The Christian Bible tells believers to turn again and again to nature to understand the essence of spiritual values. And certainly all the diverse religions of the world pay homage to the role nature plays in our humanization, our spiritual self-actualization. In her essay "Turning Slowly Nature," Diane Glancy offers this insight: "It seems to me that nature is an unsaved world. A world groaning for redemption, for release from fear, guardedness, a state of alertedness, a predatory state. Nature longs for release. Creation groans for deliverance like the humanity that inhabits it. In the biblical book of Romans (8.21) we are told, "The creation itself will be delivered from the bondage of corruption into the glorious liberty of the children of God".

As we work to redeem nature, to rescue and preserve our natural environment so that future generations may be at home here we claim our own salvation as witnesses and as custodians. Writing about integrating her Native American and European ancestry in the essay "Becoming Metis," Melissa Nelson tells us how a commitment to deep ecology was a perspective that served her even as it was the more holistic visions offered by Native traditions which provided for her a spiritual, philosophical, and political foundation from which to grow. She explains: "To indigenous peoples, the basic tenets of deep ecology are just a reinvention of very ancient principles that they have been living by for millennia before their ways were disrupted, and in many cases destroyed, by colonial forces. To learn who I am today, on this land, I live on, I've had to recover that heritage and realize a multicultural self ... By studying the process others have gone through to embrace the cultural richness of diverse backgrounds, I have come to understand the importance of decolonizing my mind." We must all decolonize our minds in Western culture to be able to think differently about nature, about the destruction humans cause.

With prophetic vision Enrique Salmon explains in "Sharing Breath" that "Cultural Survival can be measured by the degree to which cultures maintain a relationship with their bioregions.

Ecologists and conservation biologists recognized an important relationship between cultural diversity and biological diversity ... .
Cultural histories speak the language of the land. They mark the
outlines of the human/land consciousness." Our vernacular Kentucky language resonates with the richness and warmth of our land.
When we open our mouths, generations can be heard as though we
are indeed "speaking in tongues" as we embrace collective unconscious remembering our ancestors, remembering their love of the
land. It is that love which must lead us again and again to do all that
must be done to stop mountaintop removal, to recover the beauty
and function of coal without laying waste the earth. The culture
of Appalachia cannot live if our mountains are dead. We cannot
look to the hills and find strength if all we can see is a landscape
of destruction.

4

Touching the Earth

I wish to live because life has within it that which is good,
that which is beautiful, and that which is love. Therefore,
since I have known all these things. I have found them to be
reason enough and — I wish to live. Moreover, because this
is so, I wish others to live for generations and generations
and generations and generations.

(Lorraine Hansberry,
To Be Young, Gifted, and Black)

When we love the earth, we are able to love ourselves
more fully. I believe this. The ancestors taught me it was so. As a
child I loved playing in dirt, in that rich Kentucky soil, that was a
source of life. Before I understood anything about the pain and exploitation of the southern system of sharecropping, I understood that
grown-up black folks loved the land. I could stand with my grandfather Daddy Jerry and look out at fields of growing vegetables, tomatoes, corn, collards, and know that this was his handiwork. I could
see the look of pride on his face as I expressed wonder and awe at
the magic of growing things. I knew that my grandmother Baba's

backyard garden would yield beans, sweet potatoes, cabbage, and yellow squash, that she too would walk with pride among the rows and rows of growing vegetables showing us what the earth will give when tended lovingly.

From the moment of their first meeting, Native American and African people shared with one another a respect for the life-giving forces of nature, of the earth. African settlers in Florida taught the Creek Nation run-aways, the "Seminoles,' methods for rice cultivation. Native peoples taught recently arrived black folks all about the many uses of corn. (The hotwater cornbread we grew up eating came to our black southern diet from the world of the Indian.) Sharing the reverence for the earth, black and red people helped one another remember that, despite the white man's ways, the land belonged to everyone. Listen to these words attributed to Chief Seattle in 1854:

> How can you buy or sell the sky, the warmth of the land? The idea is strange to us. If we do not own the freshness of the air and the sparkle of the water, how can you buy them? Every part of this earth is sacred to my people. Every shining pine needle, every sandy shore, every mist in the dark woods, every clearing and humming insect is holy in the memory and experience of my people ... We are part of the earth and it is part of us. The perfumed flowers are our sisters; the deer, the horse, the great eagle, these are our brothers. The rocky crests, the juices in the meadows, the body heat of the pony, and man — all belong to the same family.

The sense of union and harmony with nature expressed here is echoed in testimony by black people who found that even though life in the new world was "harsh, harsh,' in relationship to the earth one could be at peace. In the oral autobiography of granny midwife Onnie Lee Logan, who lived all her life in Alabama, she talks about the richness of farm life — growing vegetables, raising chickens, and smoking meat. She reports:

> We lived a happy, comfortable life to be right outa slavery times. I didn't know nothin else but the farm so it was happy and we was happy … We couldn't do anything else but be happy. We accept the days as they come and as they were. Day by day until you couldn't say there was any great hard time. We overlooked it. We didn't think nothin about it. We just went along. We had what it takes to make a good livin and go about it.

Living in modern society, without a sense of history, it has been easy for folks to forget that black people were first and foremost a people of the land, farmers. It is easy for folks to forget that at the first part of the twentieth century, the vast majority of black folks in the United States lived in the agrarian south.

Living close to nature, black folks were able to cultivate a spirit of wonder and reverence for life. Growing food to sustain life and flowers to please the soul, they were able to make a connection with the earth that was ongoing and life-affirming. They were witnesses to beauty. In Wendell Berry's important discussion of the relationship between agriculture and human spiritual well-being, *The Unsettling of America*, he reminds us that working the land provides a location where folks can experience a sense of personal power and well-being:

> We are working well when we use ourselves as the fellow creature of the plants, animals, material, and other people we are working with. Such work is unifying, healing. It brings us home from pride and despair, and places us responsibly within the human estate. It defines us as we are: not too good to work without our bodies, but too good to work poorly or joylessly or selfishly or alone.

There has been little or no work done on the psychological impact of the "great migration' of black people from the agrarian south to the industrialized north. Toni Morrison's novel, *The Bluest Eye*, attempts to fictively document the way moving from the agrarian south to the industrialized north wounded the psyches of black folk. Estranged from a natural world, where there was time for silence and contemplation, one of the "displaced' black folks in Morrison's novel, Miss Pauline, loses her capacity to experience the sensual world around her when she leaves southern soil to live in a northern city. The south is associated in her mind with a world of sensual beauty most deeply expressed in the world of nature. Indeed, when she falls in love for the first time she can name that experience only by evoking images from nature, from an agrarian world and near wilderness of natural splendor:

> When I first seed Cholly, I want you to know it was like all the bits of color from that time down home when all us chil'ren went berry picking after a funeral and I put some in the pocket of my Sunday dress, and they mashed up and stained my hips. My whole dress was messed with purple, and it never did wash out. Not the dress nor me. I could feel that purple deep inside me. And that lemonade Mama used to make when Pap came in out of the fields. It be cool and yellowish, with seeds floating near the bottom. And that streak of green them june bugs made on the tress that night we left from down home. All of them colors was in me. Just sitting there.

Certainly, it must have been a profound blow to the collective psyche of black people to find themselves struggling to make a living in the industrial north away from the land. Industrial capitalism was not simply changing the nature of black work life, it altered the communal practices that were so central to survival in the agrarian south. And it fundamentally altered black people's relationship to the body. It is the loss of any capacity to appreciate her body, despite its flaws, Miss Pauline suffers when she moves north.

The motivation for black folks to leave the south and move north was both material and psychological. Black folks wanted to be free of the overt racial harassment that was a constant in southern life and they wanted access to material goods — to a level of material well-being that was not available in the agrarian south where white folks limited access to the spheres of economic power. Of course, they found that life in the north had its own perverse hardships, that racism was just as virulent there, that it was much harder for black people to become landowners. Without the space to grow food, to commune with nature, or to mediate the starkness of poverty with the splendor of nature, black people experienced profound depression. Working in conditions where the body was regarded solely as a tool (as in slavery), a profound estrangement occurred between mind and body. The way the body was represented became more important than the body itself. It did not matter if the body was well, only that it appeared well.

Estrangement from nature and engagement in mind/body splits made it all the more possible for black people to internalize white-supremacist assumptions about black identity. Learning contempt for blackness, southerners transplanted in the north suffered both culture shock and soul loss. Contrasting the harshness of city life with an agrarian world, the poet Waring Cuney wrote this popular poem in the 1920s, testifying to lost connection:

> She does not know her beauty
> She thinks her brown body
> has no glory.
> If she could dance naked,
> Under palm trees
> And see her image in the river

She would know.
But there are no palm trees on the street,
And dishwater gives back no images.

    For many years, and even now, generations of black folks who migrated north to escape life in the south, returned down home in search of a spiritual nourishment, a healing, that was fundamentally connected to reaffirming one's connection to nature, to a contemplative life where one could take time, sit on the porch, walk, fish, and catch lightning bugs. If we think of urban life as a location where black folks learned to accept a mind/body split that made it possible to abuse the body, we can better understand the growth of nihilism and despair in the black psyche. And we can know that when we talk about healing that psyche we must also speak about restoring our connection to the natural world.

    Wherever black folks live we can restore our relationship to the natural world by taking the time to commune with nature, to appreciate the other creatures who share this planet with humans. Even in my small New York City apartment I can pause to listen to birds sing, find a tree and watch it. We can grow plants — herbs, flowers, vegetables. Those novels by African-American writers (women and men) that talk about black migration from the agrarian south to the industrialized north describe in detail the way folks created space to grow flowers and vegetables. Although I come from country people with serious green thumbs, I have always felt that I could not garden. In the past few years, I have found that I can do it — that many gardens will grow, that I feel connected to my ancestors when I can put a meal on the table of food I grew. I especially love to plant collard greens. They are hardy, and easy to grow.

    In modern society, there is also a tendency to see no correlation between the struggle for collective black self-recovery and ecological movements that seek to restore balance to the planet by changing our relationship to nature and to natural resources. Unmindful of

our history of living harmoniously on the land, many contemporary black folks see no value in supporting ecological movements, or
see ecology and the struggle to end racism as competing concerns.
Recalling the legacy of our ancestors who knew that the way we
regard land and nature will determine the level of our self-regard,
black people must reclaim a spiritual legacy where we connect our
well-being to the well-being of the earth. This is a necessary dimension of healing. As Berry reminds us:

> Only by restoring the broken connections can we be healed.
> Connection is health. And what our society does its best to
> disguise from us is how ordinary, how commonly attain-
> able, health is. We lose our health — and create profitable
> diseases and dependencies — by failing to see the direct
> connections between living and eating, eating and work-
> ing, working and loving. In gardening, for instance, one
> works with the body to feed the body. The work, if it is
> knowledgeable, makes for excellent food. And it makes one
> hungry. The work thus makes eating both nourishing and
> joyful, not consumptive, and keeps the eater from getting
> fat and weak. This health, wholeness, is a source of delight.

Collective black self-recovery takes place when we begin
to renew our relationship to the earth, when we remember
the way of our ancestors. When the earth is sacred to us, our
bodies can also be sacred to us.

5

Reclamation and Reconciliation

Although I had been raised to think of myself as a southerner, it was not until I lived away from my native state of Kentucky that I begin to think about the geography of North and South. That thinking led me to consider the history of the African-American farmer in the United States. Coming from a long legacy of farmers, from rural America, when I left the state, I was initially consistently puzzled by the way in which black experience was named and talked about in colleges and university settings. It was always the experience of black people living in large urban cities who defined black identity. No one paid any attention to the lives of rural black folks. No matter that before the 1900s ninety percent of all black people lived in the agrarian South. In the depths of our psychohistory we have spent many years being agrarian, being at home on the earth, working the land. Cities are not our organic home. We are not an organically city people.

Even though the men and women in my family history farmed, living off the land, I was not raised to be a farmer or a farmer's wife.

My hands failed at quilting, at growing things. I could not do much
with the needle or the plow. I would never follow aunts, uncles,
nephews, and cousins into the tobacco fields. I would not work on
the loosening floor. The hard, down and dirty work of harvesting
tobacco would not determine my way of life. My destiny, the old
folks constantly told me was different. They had seen it in dreams.
In the stillness of the night they had spoken with god; the divine let
them know my fate. While they could not tell me the nature of that
fate, they were confident that it would be revealed. My elders encouraged me to accept all that was awaiting me, to claim it. Even if
claiming it meant I had to leave my home, my native place. "Jesus,"
they would tell me, "had to turn away from mother and father and
make his own way. And was it not also my destiny to follow in the
path of Jesus."

Even though I left the land, left my old grandfathers sharecropping, plowing massa's field just as though plantation culture had
never come to an end or sometimes plowing the plots of land, the
small farms that were their very own to do with as they wanted,
I was taught to see myself as a custodian of the land. Daddy Jerry
taught me to cherish land. From him I learned to see nature, our
natural environment, as a force caring for the exploited and oppressed black folk living in the culture of white supremacy. Nature
was there to teach the limitations of humankind, white and black.
Nature was there to show us god, to give us the mystery and the
promise. These were Daddy Jerry's lessons to me, as he lifted me
onto a mule, as we walked the rows and rows of planted crops
talking together.

It was sheer good fortune that I was allowed to walk hand in
hand with strong black men who cared for me body and soul, men
of the Kentucky backwoods, of the country. Men who would never
think of hurting any living thing. These black men were gentle and
full of hope. They were men who planted, who hunted, who harvested. They shared their bounty. As I take a critical look at what
black males have collectively become in this nation, defeated and

despairing, I recognize the psychic genocide that took place when black men were uprooted from their agrarian legacy to work in the industrialized North. Working the land, nurturing life, caring for crops and animals, had given black men of the past a place to dream and hope beyond race and racism, beyond oppressive and cruel white power. More often than not black females worked alongside farming black men, sometimes working in the fields (there was no money for hiring workers) but most times creating homeplace. In my grandmother's kitchen, soap was made, butter was churned, animals were skinned, crops were canned. Meat hung from the hooks in the dark pantry and potatoes were stored in baskets. Growing up, this dark place held the fruits of hard work and positive labor. It was the symbol of self-determination and survival.

There is so little written about these agrarian black folks and the culture of belonging they created. It is my destiny, my fate to remember them, to be one of the voices telling their story. We have forgotten the black farmer, both the farmer of the past, and those last remaining invisible farmers who still work the land. It has been in the interest of imperialist white supremacist capitalist patriarchy to hide and erase their story. For they are the ancestors who gave to black folk from slavery on into reconstruction an oppositional consciousness, ways to think about life that could enable one to have positive self-esteem even in the midst of harsh and brutal circumstances. Their legacy of self-determination and hard work was a living challenge to the racist stereotype that claimed blacks were lazy and unwilling to work independently without white supervision.

Black male writer Ernest Gaines recalls the spirit of these agrarian visionaries in his novel *A Gathering of Old Men*, as he also evokes the recognition that their legacy threatened those in power and as a consequence was marked for erasure. Remembering the folks who worked the land his character Johnny Paul exclaims: "They are trying to get rid of all proof that black people ever farmed this land with plows and mules — like if they had nothing from the starten but motor machines ... . Mama and Papa worked too hard in these same fields. They mama and they papa worked too hard, too hard to have that tractor just come in that graveyard and destroy all proof that they ever was." Within imperialist white supremacist capitalist culture in the United States there has been a concentrated effort to bury the history of the black farmer. Yet somewhere in deeds recorded, in court records, in oral history, and in rare existing written studies is the powerful truth of our agrarian legacy as African-Americans. In that history is also the story of racist white folks engaged in acts of terrorism chasing black folk off the land, destroying our homeplace. That story of modern colonialism is now being told. Recent front-page articles in the Lexington, Kentucky newspaper, the *Herald-Leader*, highlighted the historical assaults on black landowners. In a section titled "Residue of A Racist Past," Elliot Jaspin's article, "Left Out of History Books," tells readers that "Beginning in 1864 and continuing for about 60 years whites across the United States conducted a series of racial expulsions, driving thousands of blacks from their homes to make communities lily-white." Black farmers, working their small farmers, were often a prime target for white folks who wanted more land.

In my family, land was lost during hard times. Farming was looked down upon by the black elites active in racial uplift who had no more respect for agriculture than their affluent white counterparts. Contempt for the poor black farmer had become widespread in the latter part of the nineteenth century as black people begin to desire affluence. W. E. B. DuBois' vision of the talented tenth did not include farmers. Despite his internalized racism Booker T. Washington was the black male leader who understood the importance of land ownership, of our agrarian roots. He understood that knowing how to live off the land was one way to be self-determining. While he was misguided in thinking that white paternalism was useful and benevolent, he remains one of the historical champions of the black farmer. He understood the value and importance of land ownership, of agriculture. The elite did not favor Washington's focus on vocational training. They did not value his work with Native Americans nor his lifelong concern for the fate of poor black folk. In his autobiography, *Up from Slavery*, Washington urged black folks to choose self-reliance: "Go out and be a center, a life-giving center, as it were, to a whole community, when the opportunity comes, when you may give life where there is no life, hope where there is no hope, power where there is no power. Begin in a humble way, and work to build up institutions that will put black people on their feet." Agriculture was one arena where Washington saw black folks excelling. Working the land was one place where he could see black folks creating a culture of belonging.

In *Rebalancing the World* Carol Lee Flinders cites these characteristics of a culture of belonging — "intimate connection with the land to which one belongs, empathetic relationship to animals, self-restraint, custodial conservatism, deliberateness, balance, clarity, honesty, generosity, egalitarianism, mutuality, affinity for alternative modes of knowing, playfulness, and openness to Spirit." These core values of belonging were not taught to me by teachers and professors. Certainly in graduate school and beyond it was the culture of enterprise that mattered, what we were taught would determine our success in life. At no point in my liberal arts education was farming ever mentioned. When I first went to college and named Kentucky as my native state, laughter was often the response. Stereotypes about Kentucky, about hillbillies and the like, were the norm. No one talked about the Kentucky I knew most intimately. No one mentioned black farmers at Stanford University in my classes. Everywhere I journeyed the world of environmental activism was characterized by racial and class apartheid. In those locations no one ever assumed that black folks cared about land, about the fate of the earth.

Meanwhile in the small-town Kentucky world of my upbringing the elders were dying and the young had no interest in farming, in land. The organic gardens, the animals raised both in the country on farms and in city limits that were a way of life for my grandparents were a legacy no one wanted to preserve. And the bounty their labor brought to our impoverished and needy world was soon forgotten. Wherever I lived I made an effort to grow vegetables, even if just in pots, to garden as a tribute to the elders and the agrarian traditions they held to be sacred and as a way to hold on to those traditions. Like my maternal and paternal grandparents, I wanted to be self-reliant, to live simply. My father's father had worked land in the country, sharecropping. From him I learned much about farming and rural life. My maternal grandparents lived in city limits as though they were living in the country. They all believed in the dignity of labor. They all taught that the earth was sacred.

No one talked about the earth as our mother. They did not divide the world into the neat dualistic gendered categories that are common strategies both in reformist feminist movement and in environmental activism. The earth, they taught me, like all of nature, could be life giving but it could also threaten and take life, hence the need for respect for the power of one's natural habitat. Both grandparents owned land. Like Booker T. Washington, they understood that black folks who had their "forty acres and a mule," or even just their one acre, could sustain their lives by growing food, by creating shelter that was not mortgaged. Baba and Daddy Gus, my maternal grandparents, were radically opposed to any notion of social and racial uplift that meant black folks would lead us away from respect for the land, that would lead us to imitate the social mores of affluent whites. They understood the way white supremacy and its concomitant racial hierarchies led to the dehumanization of black life.

To them it was important to create one's own culture — a culture of belonging rooted in the earth. And in this way they shared a common belief system with that of anarchist poor white folks. Lots of poor Kentuckians, black and white, never embraced the renegade beliefs of the backwoods. But for those po' folks who did, they lived with a different set of values. And contrary to negative stereotypes those oppositional ways of thinking, those different values, were more often than not life sustaining. In *Dreaming the Dark,* feminist activist Starhawk shares this powerful insight: "When we really understand that the earth is alive, and know ourselves as part of that life, we are called to live our lives with integrity, to make our actions match our beliefs, to take responsibility for creating what we would have manifest, to do the work of healing." These were the values taught to me by my agrarian ancestors. It is their wisdom that informs my efforts to call attention to the restorative power of our relationship to nature. Collective healing for black folks in the diaspora can happen only as we remember in ways that move us to action our agrarian past.

Individual black folk who live in rural communities, who live on land, who are committed to living simply, must make our voices heard. Healing begins with self-determination in relation to the body that is the earth and the body that is our flesh. Most black people live in ways that threaten to shorten our lives, eating fast foods, suffering from illnesses that could be prevented with proper nutrition and exercise. My ancestors were chain smokers, mostly rolling their own smokes from tobacco grown locally and many of them were hard drinkers on the weekends. Yet they ate right, worked hard, and exercised every day. Most of them lived past seventy. We have yet to have movements for black self-determination that focus on our relation to nature and the role natural environments can play in resistance struggle. As the diverse histories of black farmers are uncovered, we will begin to document and learn. Many voices from the past tell us about agriculture and farming in autobiographical work that may on the surface offer no hint that there is documentation of our agrarian history contained within those pages. Anthropologist Carol Stack offers information about black farmers in *Call To Home: African Americans Reclaim the Rural South,* explaining: "After the Civil War, beginning with no capital or equity of any kind, freedmen began working to assemble parcels of land. By 1920

more than 900,000 black Americans, all but a handful of them in
the South, were classified as farm operators, representing about 20
percent of southern farmers ... One-fourth of black farmers were
true landowners, controlling a total of 15 million acres of farmland."
Stack documents the way in which black folks struggled and worked
to own land, even if that land would simply a small farm, averaging,
Stack reports "one-third the acreage of white farms."

Reading the autobiography of an African-American midwife
in the deep South whose family lived off the land and were able
to live well during hard times served as a catalyst compelling me
to think and write about growing up in rural southern culture.
Much of what we hear about that past is framed around discussion
of racist exploitation and oppression. Little is written about the
joy black folks experienced living in harmony with nature. In
her new book, *We Are The Ones We Have Been Waiting For*, Alice
Walker recalls: "I remember distinctly the joy I witnessed on the
faces of my parents and grandparents as they savored the sweet
odor of spring soil or the fresh liveliness of wind." It is because
we remember the joy that we call each other to accountability in
reclaiming that space of agency where we know we are more than
our pain, where we experience our interdependency, our oneness
with all life.

Alice Walker contends: "Looking about at the wreck and ruin
of America, which all our forced, unpaid labor over five centuries
was unable to avert, we cannot help wanting our people who have
suffered so grievously and held the faith so long, to at last experience
lives of freedom, lives of joy. And so those of us chosen by Life to
blaze different trails than the ones forced on our ancestors have explored the known universe in search of that which brings the most
peace, self-acceptance and liberation. We have found much to inspire us in Nature. In the sheer persistence and wonder of Creation
itself." Reclaiming the inspiration and intention of our ancestors
who acknowledged the sacredness of the earth, its power to stand
as witness is vital to our contemporary survival. Again and again in

slave narratives we read about black folks taking to the hills in search of freedom, moving into deep wilderness to share their sorrow with the natural habitat. We read about ways they found solace in wild things. It is no wonder that in childhood I was taught to recite scripture reminding me that nature could be an ally in all efforts to heal and renew the spirit. Listening to the words of the psalmist exclaiming: "I will lift up mine eyes until the hills from whence cometh my help."

Seeking healing I have necessarily returned to the Kentucky hills of my childhood, to familiar rural landscapes. It is impossible to live in the Kentucky of today and not feel sorrow about all that humans have done to decimate and destroy this land. And yet even as we grieve we must allow our sorrow to lead us into redemptive ecological activism. For me that takes myriad forms — most immediately acquiring land that will not be developed, renewing my commitment to living simply, to growing things. I cherish that bumper sticker that wisely reminds us "to live simply so that others may simply live." Now past the age of the fifty, I return to a Kentucky where my elderly parents live. I see the beautiful neighborhoods of my childhood, the carefully tended lawns, the amazing flower gardens making even the poorest shack a place of beauty, turned into genocidal war zones as drugs destroy the heart of the community. Addiction is not about relatedness. And so it takes us away from community, from the appropriate nurturing of mind, body, and spirit. To heal our collective spiritual body the very ground we live on must be reclaimed. Significantly in his essay "The Body and the Earth," Wendell Berry shares this vital insight: "The body cannot be whole alone. Persons cannot be whole alone. It is wrong to think about bodily health as compatible with spiritual confusion, or cultural disorder, or with polluted air and water or an impoverished soil." Our visionary agrarian ancestors understood this.

Tragically the power of dominator cultural to dehumanize more often than not takes precedence over or collective will to humanize.

Contemporary black folks who embrace victimhood as the defining ethos of their life surrender their agency. This surrender cannot be blamed on white folks. In more dire straits, slavery and the years thereafter, black folks found ways to nurture life-sustaining values. They used their imaginations. They created. We must remember that wisdom to resist falling into collective despair. We must, both individually and collectively, dare to critically examine our current relationship to the earth, to nature, to ecosystems and to local and global environments.

When I examined my relationship to the rural world that I grew up in, it was clear to me that I needed to rekindle the custodial relationship to land that was a defining characteristic of my Kentucky kin. I grew up in a rural area where many black elders owned land. Some were rewarded by white employers for faithful service with the gift of an acre or two. That was often especially the case with individual black male sharecroppers who developed co-equal bonds with white bosses. Obviously, this was not the norm, but it is meaningful to register that folk can choose to move beyond the estrangement produced by exploitation and oppression to create bonds of community. Even though black farmers were more than fifty percent of the farming population as late as 1964, by 1982 farm ownership among black southerners declined. Stack offers this explanation: "As American agriculture consolidated and shook out the many poor people in its ranks, black farms went under at six times the rate of white farms. In county after county in every southern state, land that had been in black families for generations fell into the hands of white people." And more importantly, white folks who acquired land cheap, especially land previously owned by black folks, were not willing to sell land to black folk even for high prices.

Years ago I came home to my native place to give a lecture. During the question and answer time I spoke about the white supremacy that is still pervasive when it comes to the issue of land ownership in Kentucky. Calling attention to the fact that white

Kentuckians were often willing to sell land to white folks coming from other states rather than sell land to Kentucky black folks. In some cases black folks may have come from families who for generations worked white-owned land, but when that land came up for sale their offers to buy were refused. Certainly the black Appalachian experience has always been contested by folks who either know little about Kentucky or refuse to accept the diversity of that history and the true stories of diversity in these hills. Not far from where I live in Madison County, a black man who has lived there all his life pleads with white folks to purchase land for him, and he will pay them cash. Often those rare individual black folks who purchase farmland or land in the hills find themselves paying more than their white counterparts would pay. In the old days, after slavery and reconstruction, this was called the "race tax" — "you can get it but you gotta pay more." Your paying more reassures the racist white seller that white supremacy is still the order of the day for the white folks have shown they are smarter.

When I first purchased land in the Kentucky hills, I was first a silent partner with a white male friend. We did not know whether or not the owner of the property would have been prejudiced against black folks, but we chose not to openly disclose our partnership until all transactions were completed. Many of my white friends and acquaintances who own land in the Kentucky hills are gay yet their gayness is not initially visible, and shared whiteness makes it possible for them to move into areas that remain closed to black folk because of prejudice. Liberal and progressive white folks who think it "cool" to buy land next to neighbors that are openly racist rarely understand that by doing so they are acting in collusion with the perpetuation of white supremacy. I like to imagine a time when the progressive non-black folks who own hundreds and hundreds of acres will sell small lots to black people, to diverse groups of people so that we might all live in beloved communities which honor difference. M. Scott Peck introduced his book *The Different Drum: Community Making and Peace* with the powerful insight that: "In and through community is the salvation of the world." By definition, he tells us, community is inclusive.

Writing about the issue of race in *The Hidden Wound* published in 1968 and then again in the 1988 afterword, Wendell Berry reminds us that issues of freedom and prosperity cannot be separated from "the issue of the health of the land," that "the psychic wound of racism had resulted inevitably in wounds in the land, the country itself." My own deep wounds, the traumas of my Kentucky childhood, are marked by the meeting place of family dysfunction and the disorder produced by dominator thinking and practice, the combined effect of racism, sexism, and class elitism. When I left Kentucky I hoped to leave behind the pain of these wounds. That pain stayed with me until I began to do the work of wholeness, of moving from love into greater understanding of self and community. It is love that has led me to return home, to the Kentucky hills of my childhood, where I felt the greatest sense of being one with nature, of being free. In those moments I always knew that I was more than my pain. Returning to Kentucky, doing my part to be accountable to my native place, enables me to keep a sublime hold on life.

Every day I look out at Kentucky hills. They are a constant reminder of human limitations and human possibilities. Much hurt has been done to these Kentucky hills and yet they survive. Despite devastation, and the attempts by erring humans to destroy these hills, this earth, they will remain. They will witness our demise. There is divinity here, a holy spirit that promises reconciliation.

6

To Be Whole
and Holy

Reading from a Kentucky journal written before I left my homeplace to live elsewhere, I find these words: "Troubled in mind and heart I take to the hills." The sublime happy years of my childhood were spent roaming the hills. Also in my journal I wrote: "The hills are where I am home." As a family we were isolated in the hills surrounded by nature, not another house in sight. Houses in the hollows close to ours were inhabited by poor white folk, whom we were taught were rabid racists. They were not our friends. Even if they were by chance neighborly, we were taught to mistrust their kindness. We were taught to see their friendliness as simply a gesture aimed at luring us into a trap where we would be wounded and hurt like any captured animal. No wonder then that as children we feared and yet were fascinated by white hillbillies. Individual black folk justified their anti-poor white color prejudice by saying they did not like them any more than they liked us, calling them by derogatory names like po' pecks, peckerwoods, and po' white trash. The disdain with which some black folks regarded poor white folk was definitely an inherited legacy of white supremacist hierarchies.

Privileged-class white folks looked down on the poor white folks who lived outside the law, projecting onto them many of the same negative stereotypes they used to define black people. They defined poor white people as ignorant, lazy, lawless. They talked about the broken-down cars in their yards, the trash, the way they littered their world with random objects like mouse droppings. No wonder then that most fully colonized black folks taught how and what to think by imperialist white supremacist capitalist patriarchs looked down on poor white folk, seeing them as an example of what not to be and become. Black folks were told by the white folks who dominated the poor of all colors that poor white folks were mean, cold hearted, the kind of people you had to stay away from, people who could soil and contaminate you.

I have written in the book *Class Matters* about the way in which I witnessed black teenagers mock and ridicule the one poor white female who rode the bus to school with us. They taunted her with all the negative stereotypes about poor white folks shouting loudly "She smells!" or "She stinks!" She was regarded with disdain and contempt, looked down upon the way racist white folks of all classes looked down upon black folks. This lone representative of the white working class boarded a bus daily where she often had to sit alone. If there were no seats available next to someone who was not verbally abusive (oftentimes she sat next to me), she would stand, juggling books while holding tightly to the overhead hand rail, knowing all the while that if she fell her tormentors would laugh and shout.

This persecution of an individual poor white girl by a group of black boys and girls revealed the depths of our internalized racism as colonized black people. That internalized white supremacy had taught black folks to regard any white person who would "choose" to come upon us, to be near us, near enough to touch our flesh, with contempt so strong it was akin to hatred. Such a response laid bare the reality of black self-hatred. There is no way we could collectively love ourselves and yet hate those who were most like us in habits and lifestyle.

The Kentucky hills I roamed as a child were racially integrated. Since they were outside the realm of the city, they were a location of possibility. Folk who lived there could make their own rules. In that space apart, laws could be broken and boundaries could be transgressed. There, in those lush green hills, the innate wildness of the human animal expressed itself. No wonder then that black and white in those hills feared and fascinated one another.

Not enough has been written about the psychohistory of racism in the United States, the ways in which the traumas that are a consequence of exploitation and oppression leave their mark. When I returned to Kentucky and bought acres of land in the hills, I was surprised that my six siblings (most of whom have lived in dangerous urban environments) expressed fear about living in the hills, fear of the poor white folks who live nearby. To my knowledge none of us has been wounded or assaulted by poor white folks and yet the memory of all those childhood lessons teaching us to see poor whites as the enemy made lasting imprints, marks so deep that some of my siblings say that they could not stay a night in these hills. Concurrently, African-American colleagues who teach at the college where I am a distinguished professor in residence in Appalachian Studies, a college with a long legacy of anti-racist activism, warn me, deploying the same language and stereotypes of the past, telling me that it is dangerous for me to think that it is safe to live in close proximity to "rednecks." While I do not feel afraid, I recognize that there may be some white folks who resent my presence, the white folks with confederate flags and bumper sticks that declare "heritage not hate." My response is to share with anyone who listens that the history the confederate flag evokes is one of both "heritage and hate."

Racial hatred and the racist actions it engenders are not the exclusive domain of poor whites. When I ask folks black and white who ask me whether or not I feel fear in these Kentucky hills, I counter first by asking why it is they assume that I am "safe" or more safe in the middle-class predominately white neighborhood that is my in town residence. Class prejudice is at the core of their belief that these white people are "safer" and more likely to be free of racial prejudice. In actuality, I have found white neighborhoods in all the privileged-class environments I have lived in throughout the United States, including Kentucky, to have as active a presence of racial prejudice as their poor counterparts. Significantly, those who allow that prejudice to lead them to hostile acts are in the minority no matter the class standing of the neighborhood. And in poor or privileged predominantly white environments I have found that when the few engage in active racist assaults, the many rarely take a courageous stand.

Writing about growing up in the segregated world of small-town Louisiana in the essay "Dark Waters," black male poet Yusef Komaunyaka vividly recalls the atmosphere of fear and mistrust that was the dominant force in encounters between black and white folk. Remembering, he writes: "I grew up in a climate of distrust. Blacks didn't trust whites, and it was sometimes difficult to disentangle truth from myth and folklore. For example, no black person would sell illegal, homemade liquor, but there was a white man who sold his brew to blacks. Not only did he sell 'stoopdown' under the nose of the law, but it was rumored that he doctored his corn whiskey with pinches of Red Devil lye. We believed that some among us were slowly being poisoned. This is the kind of thing that fosters mistrust ... 'There's nothing a white man won't do to keep a black man down,' they'd say: 'If he can't legally keep you in chains, he'll connive some way to keep his foot on your neck ... .' This was the folk wisdom from my community." So little has been written about the ways in which living in racial apartheid damaged the psyches of black folks, creating in some of us a pathological fear of whiteness, a fear rooted in unresolved trauma, that there is little open discussion of the way in which this psychohistory, the legacy of racialized trauma, keeps many black folks fearful of whites, convinced that all white folk have a deep-seated will to harm us. This fear and the profound mistrust it engenders is especially intense among poor folks.

Years ago when racial segregation was the norm, most folks learned about folks from a different ethnic/racial background by hearsay, that is, relying on stereotypes, gossip, and fantasy-based projections. Nowadays, mass media is the location where most folks gather information about the "other," that is, folks who are different from themselves. Unfortunately, since the culture produced in mass media often uses existing stereotypes and biases for its raw material, information about poor blacks and whites is largely negative. Yet this negative content does not constitute a closed system of thinking and being. Making the choice to look at images or read about people different from oneself, irrespective of whether those images are positive or negative, opens up the possibility that curiosity will be awakened and lead to positive contact. When my brother and I roamed the Kentucky hills meeting white "hillbillies" we found they were not all alike, that they were not all hateful. Yet the boundaries separating our two worlds precluded the possibility of friendship. Even so it may be that those early positive imprints provided foundation that enables me to meet white "hillbilly" neighbors today with the respect and openness they deserve. Just as I did not prejudge the white residents in the more privileged area where I live some of the time, I do not prejudge my white neighbors in the hills. In both places I exercise careful vigilance when needed because racism is an active systemic disorder impinging on all our lives, particularly those of people color, some of the time.

While I counted Wilma, the white female from a poor background who rode the bus with us, among my high-school cohorts and friends, I did not stay in contact with her. Of course I have wondered about her fate, pondered whether or not the persecution she experienced at the hands of young black peers on the bus marked her for life. I have wondered whether she nurtured anti-black sentiments that stayed with her from her teenage years into adulthood. Or whether she was like me, one of the ones who understood the interconnectedness of race and class early. I like to imagine that she knew the reasons she was attacked and pitied her attackers, understanding the way in which internalized self-hate fueled their prejudices. Through time, what has become clear to me is how much a cultural politics of white supremacy separates poor southern black folk from their white counterparts with whom they share a common class reality. While I felt friendship with Wilma, I did not understand in my teen years that our closeness was forged by the bonds of a shared class reality. We had both been taught to think about race but not about class.

When I left the Kentucky hills, I thought that I would be leaving behind a harsh world of white supremacy, racial hatred and prejudice, for a more enlightened environment. There was no overt awareness on my part that leaving Kentucky was also about class mobility. In the world of my upbringing where class and race converged, there was only a limited range of possibilities for a black girl from a poor and working-class background. I could be a teacher, a maid, or a housewife. If there were black college professors in our town, I did not know them. Just as I did not know that choosing to leave Kentucky to live in California and study at Stanford University was the start of a journey away from being working class, southern, country (in fact a low-class black hillbilly) to being geographically neutral. By erasing those markers I would make myself ready to engage in a class mobility that would move me up, up, and away.

Yet I was never able to truly "get away." In my mind and imagination I was always returning to the Kentucky hills, to find there a way to ground my being, a place of spiritual sustenance. This internal landscape, the world of the Kentucky hills, where I felt the deepest sense of freedom in my girlhood, was the site I returned to in my imagination to restore my soul as I lived a life in exile far away from the only place where I had felt a true sense of belonging. No wonder then that I had to return to those Kentucky hills to reclaim my sense of belonging on the earth.

When I left the Kentucky hills to attend a prestigious university out west I did not know then that more than thirty years would go by before I would return to live in the state that formed the ground of my being. Even though not a year of my life passed without my coming home to Kentucky to visit, mostly I could not imagine myself living there again. Each time I came home it seemed that so little had changed. Particularly it seemed that the harsh racial apartheid which shrouded my girlhood life in fear and rage was still unchallenged. Coming home to visit, I never went to the hills. The place I was raised as a young girl had become just another destroyed piece of earth, violated to make way for new construction. All the years I returned home to visit I sought sanctity in my parents' house and rarely ventured out. Now and then I ventured out to the porch or walked in the back yard. But I did not take to the hills. I did not want to experience again and again painful leavetaking. There was no pain when I bid farewell to the family. After visits which showed that the old Kentucky home was just as dysfunctional as always, the passion that led me home turned to a desperate will to leave and never come back. To me the family has always been that place of familiarity that holds and hurts us. Living away from home I dreamed always myself there and yet I did not think that I would ever return to live in Kentucky.

Even as years past and the years of my growing up in Kentucky served as a catalyst and a resource for much intimate autobiographical writing, I still did not consider returning to live in my home state. To return home was to come back to the pain and hurt that I had spent years of my life working to make go away. My hurt was rooted in trauma experienced in the dysfunctional family and the pain of growing up in a socially segregated world in the midst of racial apartheid. It was also hard to face the corrupt race-biased public policies that have allowed drug trafficking to devastate poor black communities. Concurrently, the flight of privileged-class black folks from our old segregated neighborhoods has made it all the more possible for these poor communities to be ravaged.

In the worlds I chose to live in away from Kentucky, I did not choose to reside in segregated black communities. These environments were more often than not places where conservative mores prevailed. I also did not choose to live in narrow-minded white communities. Instead, I chose places that were diverse, neighborhoods with ethnic, racial, religious, and sexual diversity, neighborhoods that were characterized by acceptance of difference. Even when those neighborhoods were predominately white, they were always liberal progressive locations. The environment I lived in as a teenager was firmly shaped by a social world constructed by patriarchal white supremacy and even though there was this rich sub-culture of blackness existing within and apart from this racist norm it was not a world that allowed black folk or for that matter white folks to live fully and freely. Much of the anthropological and sociological work on "return migration" (the movement of northern blacks back to southern homes) documents that many southern-born blacks long to return to the rich sub-cultures of our upbringing yet fear returning to old-style racism.

Many southern black folks long to re-capture a sense of the life lived in community with its value of relating, civility, courtesy, and mutual caretaking that most of us knew growing up. It is those ethics and values that we took with us from the South and tried to hold on to in other places. Whether living in California, Wisconsin, Florida, or Ohio, I still used my early experiences in Kentucky as the standard against which I judged the substantive quality of my life. Like many folks living in exile (this term seems appropriate because it feels to many southern blacks that sustained vicious racial segregation forced us to leave the regions of our origin), it was easier to look back at the places we left and view them in a more positive light when we were far away. Away from home I was able to look back at the world of homeplace differently, separating all that I treasure, all that I needed to cherish, from all that I dreaded and wanted to see destroyed. Like a country estate sale where all belongings are brought from a private world and are publicly exposed for everybody to gaze at them, pick them over, choosing what to reject or keep, ultimately deciding what to give away or just dump, away from home

I was able to lay bare the past and keep stored within me much that was soul nourishing. And I was able to let much unnecessary suffering and pain go.

Helped by therapy, by self-analysis, by liberating changes of mind and heart, I began to reclaim all that was precious in the Kentucky years. Appreciating that good, I could look at the elders and the folks my age who never left our Kentucky homeplace and see all the ways they were able to keep a hold onto life despite the impositions faced by the system of imperialist white supremacist capitalist patriarchy. Significantly, as I examined my past and our relationship to the environment, it was evident that generations of black folk in my homeplace had worked to maintain a profound relationship to the earth. Introducing the collection of essays *At Home On The Earth: Becoming Native To Our Place*, editor David Barnhill begins with the declaration: "Our relationship to the earth is radical: it lies at the root of our consciousness and our culture and of any sense of a rich life and right livelihood." Before our contemporary concern with the earth gained a hearing, similar sentiments were often expressed by one of the most famous black. environmentalists who has lived on the planet, George Washington Carver, who was fond of saying: "What is money when I have all the earth?" Carver wanted everyone, but especially black folks, to engage in careful husbandry of the earth.

Engaged with issues of sustainability before these concerns were popular, he continually worked to teach reverence for the earth. Understanding the way that refusal to care rightly for the earth was linked to a willingness on the part of humans to exploit and dehumanize one another, Carver spent a lifetime working to demonstrate the life-enhancing relationship humans could have with the earth, with plants. Despite his efforts many southern black folks felt that progress could be theirs only if they turned away from their agrarian roots. The end of slavery, the mass migration of agrarian southern black folks to northern cities, the demise of a collective presence of black farmers, all produced grave silences about our relationship to the earth. Breaking that silence has been crucial for individual progressive southern black folks engaged in decolonizing our minds, especially those of us who are choosing to reclaim our legacy as stewards of the earth.

Erasing the agrarian past wherein black folks worked the land, sustained our lives by growing and tending crops, was a way to deny that there were any aspects of life in the white supremacist South that was positive. At the end of nineteenth century the cultural myths that made freedom synonymous with materialism necessarily denied the dignity of any agrarian-based lifestyle. Black folks who embraced this version of the American dream were as eager as their white counterparts to leave behind an agrarian past to seek freedom in the industrialized North. The life of the black farmer was one of hard work often without substantial material reward.

Black folks (like my maternal and paternal grandparents) who felt working the land rooted them in a spiritual foundation that was mystical had no visibility in the movement for racial uplift that privileged material success above all else. By the time George Washington Carver passed away, there were no visible black leaders telling black folks that farming was a right livelihood that in his words would "unlock the golden door of freedom to our people." Carver believed that black folks could gain self-determination and self-sufficiency by living in harmony with the earth even as his transcendent vision encompassed all people. To Carver, maintaining a caring relationship to the earth, to nature was a means to have union with the divine. Time and time again he told listeners: "Nothing is more beautiful than the woods before sunrise. At no other time have I so sharp an understanding of what God means to do with me." This spiritual bond with the earth is one of the many counterhegemonic beliefs that sustained exploited and oppressed black folks during the years of slavery and reconstruction. Indeed, experiencing the divine through union with nature was a way to transcend the imposed belief that skin color and race were the most important aspects of one's identity. Leaving a rural past many black folks began to feel estranged from our southern roots, from nature. This estrangement meant that the organic spiritual renewal generated by direct engagement with the natural world was no longer a given in the daily life of ordinary black folks.

Growing up in Kentucky I knew as a child that there was a tremendous tension between those black folks who lived in the country and those who lived in the city. As a young girl I knew there were conflicts between these two Kentucky cultures — the world of the hills, the backwoods, the country, and the world of the city, civilization, law, and order. One of the most traumatic experiences of my early childhood was the movement of our family from the country to the city. In my child's mind rural life was synonymous with belonging in nature, freedom, adventure, safety; city life was about containment, restricted movement, an overdeveloped dangerous landscape. The fearlessness and awe I experienced as child belonging in nature imbued me with a power and confidence I soon lost in the city where I felt invisible, powerless, and lost. In the isolated, underpopulated, rural environment of the Kentucky hills, there had been no persistent sense of threat or danger — no need for a child to be endlessly told to be careful, to always be on guard. In the world of the city danger was everywhere. Interviewed in the magazine *Kentucky Living*, mystery writer Sue Grafton recalls her childhood in Kentucky, testifying that she continues to "value the simplicity of the world I grew in," remembering that "in those days, the world was much more innocent." She acknowledges: "I feel fortunate that I was able to have as much freedom as I did." Experiencing freedom in nature during girlhood was fundamentally empowering.

My childhood heart broke when I had to leave the country where I felt safe, the country of quiet slow days, no crowds, and a stillness never felt in a noisy small city full of fast-paced movement and strangers. Living away from a renewing natural world I felt a deep sense of soul loss that was traumatic. Away from rural Kentucky life I was taught that to be really self-actualized, to become someone of value, I would need to leave my Kentucky roots. At the end of my teenage years, my next big move was to board a bus, then a plane that would take me to California, as far from Kentucky as a girl could go, a black girl with few resources. Leaving home I fulfilled the expectations of those who had taught me to believe that I should leave Kentucky and become a better person, be born again. Leaving Kentucky triggered the underlying feelings of brokenheartedness that had surfaced during that initial move from country to city. In all the places I journeyed to in an effort to become that "better" human being away from my Kentucky home, I confronted a culture of narcissism, one in which spiritual beliefs and ethical values had very little meaning for most folk. I longed to find in those places the values that I had learned in my growing-up years. Simple values had grounded my sheltered life. Taught first and foremost to be a person of integrity, to love one's neighbor as oneself, to be loyal and to live at home on the earth, I did not know how best to live in a world where those values had no meaning.

My decision to return to Kentucky to live was rooted in a growing awareness that much of what I did not like about my native state (the persistence of a cruel and violent racism in daily life and sustained patriarchal assumptions) was more and more becoming the norm everywhere. Concurrently, remnants of all that I cherished in my childhood years were still present among my Kentucky family and community. Though old and frail my parents were and are still hanging on to life. Coming back to live in Kentucky affords me the opportunity to spend time with them during the last years of their days on earth. My father describes these years as a time when "we are going down the mountain." In Barbara Kingsolver's book *Animal, Vegetable, Miracle*, she shares that her family moves from Arizona to rural Kentucky because of many "conventional reasons for relocation" but an important one being extended family. She explains "Returning now would allow my kids more than just a hit-and-run, holiday acquaintance with grandparents and cousins. In my adult life I'd hardly shared a phone book with anyone else using my last name.

Now I could spend Memorial Day decorating my ancestors' graves with peonies from my backyards. Tucson had opened my eyes to the world and given me a writing career and legions of friends and a taste for the sensory extravagance of red hot chilies and five-alarm sunsets. But after twenty years in the desert, I'd been called home." This call to home comes at a time when many of us are ready to truly slow down and settle down.

Like many folk returning to small-town southern roots, one of the most immediate experiences that calls us is the slowing down of everything. In childhood my siblings and I often hated the languid slow pace of everyday life once chores were done. Then, we wanted action, movement, the possibility of something happening. We were not interested in sitting still. Seeking a place of spiritual grounding from Buddhist and Christian practice in my life today, I experience stillness as a path to divine mind. Now, I want to be still. I want to live in that experience of knowing unity with the divine in stillness.

Returning to the Kentucky landscape of my childhood and most importantly to the hills, I am able to reclaim a sublime understanding that living in harmony with the earth renews the spirit. Coming home to live in Kentucky was for me a journey back to a place where I felt I belonged. But it was also returning to a place that I felt needed me and my resources, a place where I as a citizen could be in community with other folk seeking to revive and renew our local environment, seeking to have fidelity to a place. Living engagement with both a specific place and the issue of sustainability, we know and understand that we are living lives of interdependence. Our thinking and our actions are constantly informed by what Wendell Berry defines as "an ecological intelligence: a sense of the impossibility of acting or living alone or solely in one's own behalf, and this rests in turn upon a sense of the order upon which any life depends and of the proprieties of place within that order." To simply return to agrarian locations without coming full circle and letting go of manners and mores taken from living in small and large cities, we have little to offer the places we return to. Hence we return to the unforgettable homeplaces of our past with a vital sense of covenant and commitment.

In the hills of my grown-up life I have an opportunity to live beyond the narrow prejudicial ways in which I was taught to see poor white folk. I have to work at appreciating the trailers dotting the hillsides that look to me like giant tin cans. Seeing beyond the surface, looking at the care neighbors bring as they work to turning these structures into homes, I feel empathy and solidarity. Writing about the meaning of community in *The Different Drum: Community Making and Peace*, M. Scottt Peck shares that true community is always "realistic." To be real in community differences must be acknowledged and embraced. Peck explains: "Because a community includes members with many different points of view and the freedom to express them, it comes to appreciate the whole of a situation ... An important aspect of the realism of community deserves mention: humility. While rugged individualism predisposes one to arrogance, the 'soft' individualism of community leads to humility. Begin to appreciate each other's gifts, and you begin to appreciate your own limitations. Witness others share their brokenness, and you will become able to accept your own inadequacy and imperfection. Be fully aware of human variety, and you will recognize the interdependency of humanity." These insights come to those of us who truly embrace community, seeking to live in fellowship with the world around us. Sadly, accepting human variety means that we must also find a way to positively connect with folks who express prejudicial feeling, even hatred. Committed to building community we are called by a covenant of love to extend fellowship even when we confront rejection. We are not called to make peace with abuse but we are called to be peacemakers.

Of course we cannot truly be peacemakers if we have not found peace within. For many folk who are disenfranchised and indigent in the Kentucky hills peace comes from that sense of belonging on this particular place on earth, our hills and mountains. This peace must be protected when we live in a culture that is daily warring against us. Just as mountaintop removal destroys and decimates sacred ground, our souls are assaulted by imperialist white supremacist capitalist patriarchy. The colonization of our Kentucky earth is as vicious as is the assault on what tourist-minded advertisers evoke with their slogan Kentucky's "unbridled spirit." When we work to protect our community as well as the earth which is our witness, the ground on which we stand, we create the conditions for harmony, fellowship, peace. Those of us who left Kentucky ground to plant the seeds of our being elsewhere return to the bluegrass state because we realized, as we made lives elsewhere, that this was the landscape, the earth, that most nourished and nourishes our spirit.

The insistence that the best and brightest of the Kentucky hillbilly country folk need to move elsewhere to become fully self-realized comes with no insistence that this elite group should return to their homeplace. Consequently a major human resource is taken away from this state. Symbolically, this is not unlike mountaintop removal. Making the decision to move back to Kentucky, I was stunned when many city-based friends and acquaintances expressed concern that I was headed to the backwoods, to a place where I could not be truly myself, truly alive. It surprised me that so many of the negative stereotypes about life in Kentucky are as fixed in our national geographic imagination as they were when I first left the state years ago.

Living in New York City and feeling as though I did not belong, I begin an inward search for place. More than thirty years have gone by since my leaving Kentucky. Hence my inward search did not begin with my home state. This seeking led to a psychic archaeological dig where I plundered the depths of my being to see when and where did I feel a sense of belonging, when and where did I feel at home in the universe. That searching led me home to Kentucky. But I could not just be anyplace in my home state. I chose a small progressive town where the me of me could live freely. In the essay "Local Matters," Scott Russell Sanders explains the importance of finding a place of belonging: "It is rare for any of us, by deliberate

choice, to sit still and weave ourselves into a place, so that we know
the wildflowers and rocks and politicians, so that we recognize faces
wherever we turn, so that we feel a bond with everything in sight.
The challenge, these days, is to be somewhere as opposed to nowhere, actually to belong to some particular place, invest oneself in
it, draw strength and courage from it, to dwell not simply in a career
or a bank account but in a community …Once you commit yourself
to a place, you begin to share responsibility for what happens there."
Even though I wanted to belong in the many places I lived away
from Kentucky, I never committed. There was always the possibility
that I would be moving on, starting over.

Returning to Kentucky, making my life in a small town, I knew
that this was the end of my journey in search of home. Sanders
shares in the essay "Settling Down" that it took him "half a lifetime
of searching to realize that the likeliest path to the ultimate ground
leads through my local ground. And he means the land, weather,
seasons, plants, animals. Boldly he declares, "I cannot have a spiritual
center without having a geographical one. I cannot live a grounded
life without being grounded in a place." Connecting homeplace to
spiritual peace, he reminds us that "in belonging to a landscape,
one feels a rightness, at-homeness, a knitting of self and world."
Eloquently these words express my feelings about being here, about
my house on the hill and the acres around it that I know will be forever green, recovered from hilltop removal, no sub-divisions.

I am called to use my resources not only to recover and protect
damaged green space but to engage in a process of hilltop healing.
Although I come from a long line of Kentucky country folk, farming women and men, I am having to learn my stewardship. I do not
have the longed-for green thumb but with a little (more than a little)
help from my community, I am doing the work of self-healing, of
earth healing, of reveling in this piecing together of my world in
such a way that I can be whole and holy.

7

Again — Segregation Must End

No doubt every writer of essays has one or two that give them pause, make them think again and again, wondering where did that come from? It is usually impossible to explain to folks who are not writers that ideas, words, the whole essay itself may come from a place of mystery, emerging from the deep, deep unconscious surfacing, so that even the writer is awed by what appears. Writing then is revelation. It calls up and stirs up. It illuminates. Among my essays, one that really shook me up and moved me is the essay "Representations of Whiteness in the Black Imagination" (see Chapter 8, this volume). In this essay I wanted to talk about the psychological traumas racism causes. In particular I wanted to write about how black folk living in the midst of racial apartheid come to fear whiteness, come to see it as something terrible. When the civil rights struggle first brought national attention to the issue of racial integration, individual racist white folks would often share their "intimate" knowledge of black folks by telling the public that the "colored people like to keep to themselves." Yet no one ever raised the issue of trauma, that maybe black folks stayed together and wanted to stay away from white folks because of the suffering white folks caused us to feel through unrelenting exploitation and oppression. The monstrous way in which white racists inflicted and continue to inflict pain and suffering on black people will never be fully recorded or acknowledged. Contemporary, big, beautiful, expensive coffee table books which bring stylish images of brutal lynchings into our homes, making it appear as though these past atrocities are just that, something over and done with, deny that these horrific images documenting the vicious hateful attacks on helpless black bodies by powerful white bodies carry with them a legacy of trauma that has not passed, that has not gone away. Our nation is capable of acknowledging that Jews who were nowhere near the German holocaust, whose relatives, friends, and acquaintances were murdered and slaughtered, suffer post-traumatic stress disorder, fear of the "German" other, fear of bonding outside one's group, and at times the crippling fear that it will happen again. But it has only been in very recent years that there has been a willingness on the part of a very minority of thinkers in the psychological community and beyond, to acknowledge that black people who witness grievous racist exploitation and oppression are traumatized. And even when incidents are over that the victims experience post-traumatic stress.

Fear of being victimized by racist abuse has long kept black folks confined in segregated neighborhoods and social relations despite legalized anti-discrimination laws and accepted racial integration. Growing up in racial segregation I felt "safe" in our all black neighborhoods. White people were represented as a danger, especially white males. Every black girl in our segregated neighborhoods knew that we had to be careful not to have any interaction with white males for they were most likely seeking to violate us in some way. While sexual violation was the dreaded form of white male racist assault, it was also clear that white folks, often acting on a whim, humiliated and shamed black folks, whether through aggressive verbal abuse (calling us by ugly racist epithets) or blatant physical assault. In the days of legal racial segregation, no black person could defend themselves against the violence of a white person without suffering severe reprisals. Consequently, black children living in racial apartheid were systematically socialized to fear white folks and to stay away from them.

Even though we lived in segregated neighborhoods, there were a few black folks in our town who lived near white folks. Our mother's mother Sarah Oldham lived in a white neighborhood. To visit her house we had to walk through neighborhoods occupied by racist white folks who taunted and jeered at us. Needless to say, as children, walking through these neighborhoods was frightening and stressful. Even if we passed the homes of white folks sitting on the porch who were friendly, we had been socialized to see their friendliness as a lure, setting a trap wherein we would be caught, where we would be helpless and hurt. Taught to be critically vigilant in relation to white people, we were not taught to see all white people as "bad." We were taught that there were good and kind white people, but that they were rare. Meeting the adversity caused by white supremacist aggression early in life helped many of us have post-traumatic growth. Were this not the case, individual black people would never have acquired the skill to live harmoniously among white people. Still many black people suffer post-traumatic stress disorder as a consequence of sustained racist exploitation and oppression. More than not that pain is usually ignored in our culture.

There is no psychological practice that specifically focuses on recovery from racist victimization. Indeed, our society has moved in the opposite direction. Many people in our nation, especially white people, believe that racism has ended. Consequently, when black people attempt to give voice to the pain of racist victimization we are likely to be accused of playing the "race" card. And there are few if any public spaces where black folks can express fear of whiteness, be it engendered by rational or irrational states of mind. However, white fear of blackness gains a constant hearing. And psychological research indicates that a great majority of white Americans respond negatively to images of blackness. In Jonathan Haidt's book, *The Happiness Hypothesis: Finding Modern Truth in Ancient Wisdom*, he calls attention to this implicit prejudice: "Researchers have found that Americans of all ages, classes, and political affiliations react with a flash of negativity to black faces or to other images and words associated with African-American culture." White Americans who see themselves as not prejudiced usually shared these same reactions. The history of racial apartheid from slavery's end to the present day has focused on the issue of integration, particularly social integration, housing and interpersonal familial relations. Today in our culture the workforce is racially integrated, people of color (especially black folk) and white folk work alongside one another, may even share lunch, but rarely does this racial integration carry over into life beyond the job. Mostly white and black in our nation live segregated.

Studies of race and real estate show housing to be an arena where racial discrimination continues to be the norm. In *Words That Wound*, law professors Mari Matsuda and Charles Lawrence discuss the reality that racism shapes "suburban geography," stating that "while residential segregation decreases for most racial and ethnic groups with additional education, income, and occupational status, this does not hold true for African-Americans." In his work on race and the issue of housing, political scientist Andrew Hacker calls attention to the fact that even liberal and progressive white folks are concerned when it seems as though more black homeowners are moving into what they perceive to be "their" neighborhoods. In his book *Two Nations*, he emphasizes that one black family may find acceptance in a predominately white neighborhood but more than one is seen as threatening. Of course most white homeowners insist that the issue is not racial prejudice but rather economics; they are concerned that their property will not rise in value. Given the system of white supremacy, the blacker the neighborhood, the more likely it is that the property therein will be deemed less valuable by property appraisers, who are usually white.

Nowadays, in real estate circles, sellers and buyers alike talk about racial economic zoning in housing. Since on the average white families make more money than people of color/black people, some neighborhoods will automatically be all white because of high prices. Certainly, in the small predominantly white Kentucky town of Berea where I reside there has been no history of racial segregation in housing. Instead the town's founder John Fee and its citizens were committed to the project of ending racism, of ending segregated housing. During the early days of the towns development in the late nineteen hundreds, citizens had to sign a covenant reinforcing this commitment, one that publicly declared their willingness to live next to a white or black neighbor. When I first moved to Berea, I, like other black folks before me wanted to see the "black neighborhoods." I was astounded when I was repeatedly told that Berea really did not have black neighborhoods, that because of its history of anti-discrimination in housing individual black people live wherever they desire. This is one of the aspects of Berea that continues to make the town a swell place to live.

Unfortunately in recent years the building of exorbitantly priced homes has created segregated all-white neighborhoods where the presence of people of color is often not welcomed and where there are residents who would not choose to live among black people in racially integrated subdivisions. Most, if not all, of these segregated white communities are new developments and sometimes potential homeowners must be vetted by associations and boards before they can purchase in these locations. This is a perfect setup for discrimination to occur. The underlying principle of many new housing developments, particularly those that are gated, is the notion of exclusion and exclusivity, keeping undesirable elements out, which frequently means people of the wrong class or color. Of course most of the residents in these communities will argue that their choice of housing is not influenced by racial prejudice, because they are not racists, but rather by a desire for comfort and safety.

Writing about the fact that most white people have been socialized to remain ignorant of the way racism affects their lives, the way they collude in maintaining racism and with it segregation, Peggy McIntosh contends: "As a white person, I realized I had been taught about racism as something which puts others at a disadvantage, but had been taught not to see one of its corollary aspects, white privilege which puts me at an advantage ... In my class and place, I did not recognize myself as a racist because I was taught to see racism only in individual acts of meanness by members of my group, never in invisible systems conferring unsought racial dominance on my group from birth." The average white American never thinks about the issue of supporting racial discrimination when it comes to housing, never questions their choice to live in segregated white communities.

In the aftermath of the sixties' civil rights struggle there were liberal white and black individuals who expressed concern about segregation in housing and actively chose to try to create diverse and beloved communities. Focus on desegregation without anti-racist thought and behavior did not help create a safe context wherein bonding across racial difference could be seen as both necessary for progress and appealing. When I left my native place and the racial apartheid that characterized all arenas of social life, especially housing, I did not want to live in white neighborhoods: I wanted to live among diverse groups of people who were interested in creating anti-racist integrated environments. Migrating to the west coast, attending Stanford University, I found myself in an environment with like-minded folks. The anti-racist energy of the sixties had impacted all or us. We wanted to create communities of love and hope. Those feelings coincided with notions of living simply. Living off the land, living with less, was deemed vital to the survival of the planet. In those days most of us either lived in very small spaces or in big houses with lots of other people. In keeping with this spirit of democracy and anti-discrimination in housing, some folks were opposed to the idea of property ownership and private property. Yet all our counterculture lifestyles did not change the core of imperialist white supremacy capitalist patriarchy. Needless to say, the alternative values and habits of being cultivated during a period of national revolt against dominator thinking and dominator culture did not last.

While many sixties' and seventies' radicals and liberals maintained progressive anti-war sentiments and ideas about gender equality, they became more conventionally conservative around issues of housing and property. As this group of people aged, many inheriting property and wealth from the coffers of more conservative elders, they became more fiscally conservative. And when the issue was housing and real estate, patterns of discrimination and segregation were reinforced even though racial integration had become more of an accepted norm. Owning property during the real estate boom of the eighties and nineties was one of the quickest ways for individuals to acquire unearned profit and in some cases wealth. Groups of liberal and progressive whites who had been at one time pro-active in the struggle to create integrated housing became more comfortable with being in all-white or predominately white communities.

Living in major cities and watching gentrification in real estate, any observer could witness a process wherein groups of more liberal whites would purchase housing in neighborhoods that were peopled primarily or solely by people of color/black people. The "cool" white folks would declare their movement into these neighborhoods was a gesture of solidarity, of openness to diversity. Yet more often than not their presence usually raised prices and increased real estates taxes. Often they were coming from privileged classes. Rather than adding to diversity, their presence usually pushed out the underprivileged colored folks. And even when poor people own property in areas that more affluent groups gentrify, the underprivileged are rarely able to sell their homes for big profits and move elsewhere.

The rising cost of housing makes it impossible for them to reap rewards if they sell and simultaneously buy a better or even similar home to the one they would leave behind. No matter the income of people of color, especially black people, our growing presence in any neighborhood does not lead to any substantial increase in the value of property. While there are many places where an influx of black residents led to white flight, there is no case where the presence of those black newcomers led to a substantial increase in property value.

Neighborhoods in the various parts of the United States that I have called home often started out racially integrated, ethnically diverse, but as the presence of white newcomers from privileged classes increased they became whiter and whiter. When I moved to New York City I chose to live in the West Village because of its diversity of race, class, and sexual orientation. However, as property values increased, not only did the culture of privileged-class wealthy whiteness become dominant, those of us deemed "different" were and are increasingly seen as undesirable interlopers. Covert expressions of white supremacy are difficult for most people to see. This seems to be especially the case when real estate is the issue. Individual white consumers who see themselves as anti-racist rarely question their choice to live in all white neighborhoods. They do not question why it is that they feel more comfortable living solely among other whites people, even if they do not have much in common. Most black people will justify their choice of segregated neighborhoods by calling attention to cheaper prices and or a refusal to live in fear, to live among folks who hate you and threaten to harm you.

Although I am happiest living in neighborhoods with diverse residents, I know that when anti-racist sentiments rule the day the color of one's neighbors does not matter. Growing up in our Kentucky town, the intense racial apartheid and the fear it engendered among black folk, the fear that we would be harmed by whites if we ventured into their neighborhoods, led me to leave my native place. I did not want to live my life in that fear, my movement, as well as all social relations, circumscribed by it. I did want to see every white person as a potential enemy. Living most of my life away from Kentucky in progressive states California and New York City, in neighborhoods where racial integration was the desired and accepted norm, I would return home and find the same old racist boundaries operating as though nothing had changed. And even when individual black folks ventured out of our segregated worlds into predominantly white areas, they faced hostility and indifference from white neighbors. Tragically, while certain social norms created by racism and white supremacy remained intact in Kentucky, they were slowly coming to be the norm in other places.

While segregation is no longer a legally imposed norm anywhere, in most places separatism is the unspoken but accepted rule. And since many black people self-segregate, they unwittingly collude with racist whites in maintaining racial segregation in housing. Speaking about the need for citizens of this nation to continue to see segregation as a political issue that must be fully addressed if racism is ever to truly end in 1995, Colin Powell made this insightful comment: "We are a nation of unlimited opportunity and serious unresolved social ills; and we are all in it together. Racial resegregation can only lead to social disintegration. Far better to resume the dream of Martin Luther King, Jr.: to build a nation where whites and blacks sit side by side at the table of brotherhood." Liberal identity politics, though often formed to serve as a basis for civil rights organization and protest, brought with it an emphasis on maintaining allegiance to one's own racial, ethnic, or social group that was at times more akin to white supremacist thinking about staying with one's own group. It created a paradox with people of color on one hand insisting that it was vital that racism end and on the other insisting on the primacy of allegiance to one's race. To a grave extent people of color who self-segregate are in collusion with the very forces of racism and white supremacy they claim they would like to see come to an end.

Racism will never end as long as the color of anyone's skin is the foundation of their identity.

When we bond on the basis of shared anti-racism, skin color is placed in its proper perspective. It becomes simply another aspect of a person's identity, not the only important aspect. Like all black people raised in a segregated world who have experienced racist assault, however relative, for the first thirty years of my life, I was more comfortable in settings where black people were in the majority. As I learned more about the structure of racism and white supremacy, I challenged myself to critically examine notions that I was "safer" in all black settings, more understood, or automatically shared a common sensibility. I wanted to have congruency between my anti-racist belief that to end racism we would all need to stop overvaluing race and the actions I might take regarding race in everyday life. I had to acknowledge that being among black people victimized by internalized racism could be just as dangerous as being among racist whites. I had to acknowledge that the ideal experience was to live among groups of people committed to living an anti-racist life. When being militantly anti-racist is a basis for bonding we are constantly called to look beyond skin color and acknowledge the content of our character.

Choosing to return to my native place, to come home to Kentucky, led me to examine anew my commitment to ending racism. The choice to return to Kentucky was never a consideration for me until I came to give a lecture at Berea College. Certainly, I have never felt and to this day do not feel that I could return to my hometown because so much racial segregation and racial separatism remains the norm there. Ironically, even though I grew up less than four hours away from Berea, Kentucky, I had never heard about the place or its history. More than ten years ago the women's studies program here invited me to come and lecture. When I accepted the invitation, I was sent information about the college and its history. Here was a town in the South, in Kentucky, that was designed around the principles of anti-racism. And that design was made manifest in 1855. The visionary founder of both the town and college, John Fee, was a white male abolitionist Christian who wholeheartedly believed that "we are all people of one blood." Long before it was "cool" to talk about race as a social construct, long before it was scientifically proven that there really is no genetic basis for racial difference, long before public knowledge of organ transplants showed the public that the internal workings of the human body are fundamentally the same irrespective of race, long before Martin Luther King preached the importance of building beloved community, Fee and his supporters were doing just that, creating an anti-racist utopian environment where white and black people could live together in peace and harmony.

Awed by Berea's history, by the prophetic vision of John Fee, a vision that with the heartfelt cooperation of like-minded folk he was able to realize, I wanted to participate in the contemporary maintaining of his radical legacy. Like many people before me, I was awed by the fact that Fee created a need-based college with no tuition, where black and white, women and men, could come together as equals and learn and live together both at the college and beyond in community. The anti-racist credo that was both the philosophical and spiritual foundation "we are all people of one blood" serves as a constant reminder of this legacy. Tragically, Fee's vision worked so well that white supremacist politicians intervened, making it legally impossible for the college to continue its inter-racial project. For many years on into the age of civil rights the anti-racist progressive vision of beloved community was suppressed in Berea. But the power of that radical vision was sustained and like a seed planted in the earth it continues to bear fruit despite the years in which that vision was betrayed and undermined. As a consequence the radical legacy of anti-racist beloved community is not nearly as fully realized as it was in Fee's time today, yet there are still enough people who remained committed to ending racism and creating beloved community to provide a sustaining home.

Coming to lecture at Berea College years ago, I was impressed with both the college and the town. Again like others before me, I wanted to come and work at the college and made that desire known. When I first lectured at the college, I spoke openly about the intense racial apartheid that was part of my Kentucky childhood. I talked about the relationship between racism and real estate, calling attention to the fact that to this day many racist white folks will not sell land to black folks, even if those folks have been workers on the land for generations. I called attention to the fact that white people from all over the nation purchase land in Kentucky and do not face the discrimination that still excludes many black Kentucky folks from land ownership. During that lecture I confessed that the persistence of cruel racist practices in everyday life made it impossible for me to imagine living in Kentucky. At that time my audience was largely made up of progressive people they were eager to call attention to the reality that Berea had a different history from the one I was speaking about. And certainly the people I met during my first visit, most of whom were white, seemed as committed to the creation of beloved community as John Fee had been. During that very first visit I decided to come home to Kentucky, now that I had found a progressive place in my home state.

To be able to return to my native place, to live and work in a small town in Kentucky that was progressive was for me a miracle. Sadly during the years where racist aggression led to a shift in the progressive anti-racist public policy of the town, many black residents left. Even though the vision of inter-racial community is still alive and flourishing in Berea, there is not a large population of adult black people. The college draws younger black people but the town does not have enough work opportunities to attract many people of color newcomers. Berea is still a small town. Many of the older white and black residents (though fewer in number) remember the way life was "back in the day" when all neighborhoods in the town were racially integrated. Today the town is predominately white. Even so, the black presence is visible in the downtown. And it certainly affirms me to see other folks of my same color. Since the presence of one black person alone does not fully integrate or change the white supremacist norm, it is important that everyone in our community choose anti-racism. Fortunately a significant number of the white residents in Berea remain committed to anti-racism in daily life. This does not mean that the college and the town are doing all that can be done to realize anew the vision of inter-racial living, of beloved community. It does mean that the foundation for such community is in place.

Given Berea's history I felt none of the anxiety engendered by awareness of racial discrimination that I had felt in New York and California. Certainly, had I been looking for housing in most places in Kentucky I would have maintained critical vigilance about the continued practice of racial discrimination in real estate. Given all the arenas of life in our nation that have become racially integrated, the workplace, stores, clubs, restaurants etc., places where helpful laws make discrimination difficult, housing remains one of the locations where it is practically impossible to "prove" discrimination is taking place and where the politics of white supremacy, racial exclusion, and segregation remain more a norm. And that is especially true in places where gated communities, condo associations, and co-opts have the right to interview candidates and turn them down for multiple reasons.

As a single black woman looking to purchase housing in New York City, supposedly one of the more progressive places in our nation, I would be told confidentially that it was not that different boards did not approve of me but that they were concerned about the people who might come to visit me (i.e., black males). In one case I was told to say nothing about my black male partner when being interviewed. I explained that I could not do this because it would not be fair to him to place him in a circumstance where he would be continually regarded with suspicion and hostility. I found it highly ironic that I confronted racial discrimination in New York City, decades after the civil rights movement. In this amazing city of diversity the fierce racial stereotyping and racial segregation that had led me to leave Kentucky were alive and well and thriving.

The conditions of racial discrimination that I had found so unjust in my growing-up years are fast becoming the norm everywhere in our nation. No wonder then that I felt especially lucky to find a small town in Kentucky where I did not need to fear that I would encounter racial separatism when looking for housing. Prophetically, John Fee's understanding that actively resisting the formation of segregated neighborhoods, making it rewarding for folks to choose integration, would lead to an end to racism in housing has proved true. No doubt he would be dismayed to see the way in which the development of new overpriced sub-divisions is leading to the very racial segregation and exclusivity that he and like-minded allies worked so hard to change.

Although I live in a predominately white neighborhood in Berea, as the town is not as racially integrated as the college, I am never afraid. While I do not know most of my neighbors intimately, it matters to me that many of the white folks who surround me are committed to ending racism. And when it comes to housing, they show that commitment by their refusal to discriminate or segregate. The folks I love, who are white, are as passionate about challenging dominator culture and ending racism as I am. They are open to being challenged. This is especially true of radical friends, gay and straight, who live outside the town, in the hills. When I question them about why they accept living on acres of land surrounded only by other white people, some of whom are white supremacists, they do give any meaningful response. They do not follow the path paved by John Fee and decide that they might need to sale or give and acre of land to people of color as a radical contestation of white supremacy. Over the years as I have bought and sold property, I have seen that irrespective of their political stance, whether to the right or the left, when it comes to the issue of land and home ownership most folks are conservative, most folks are not willing to make of their lives and their lifestyles a living practice that challenges racism. In an effort to ensure that my life practices offer an alternative vision, one that daily challenges discrimination in housing, I have bought homes to share with others, even to give away. So far the black folks I have invited to come and live in our community are reluctant to live here. Family members who are struggling economically elsewhere, whose hard-earned income goes mainly to housing, tell me things like "small towns are boring" or "there are not enough black people in Berea." One of my sisters who lives in a predominately black city in the United States shared that while she likes coming to Berea to visit, she just does not want to live where "there are not enough black people." Her comment led me to ponder "how many black folk constitute enough." And I posed this question to her: "If you have the opportunity to live in a predominately white community with a relatively small black population wherein the black and white folks who dwell there are anti-racist and caring to one another or you have the choice to live in a predominately black city where many black folks have internalized racism and are not caring towards self and others, where the small number of whites are by and large racist and/or afraid of black people, which would be the better place to live?" I raise the question with her and with other black folks of whether or not it is better to live in a community with five healthy black neighbors than fifty unhealthy black folks. This is one case where identity politics leads us astray, leads us to surrender optimal well-being for a false sense of safety that is ultimately based on white supremacist notions that we are safer and more at home with our "own kind."

When black folks no longer suffer from internalized racism, we know that we can love blackness and embrace racial integration, that cultural allegiance need not blind us to the need to recognize and live beyond the artificial boundaries set by racist notions of race. As long as black people behave as though we can only be safe in segregated spaces, white supremacy is reinforced. Today many black people feel that white supremacy is so powerful that it can not be effectively challenged. Concurrently, they believe that white people who cling to racist beliefs (sometimes unconsciously) will never change. Since so many white people deny that race is one of the primary ways in which power, unearned privilege, and material well-being are bestowed on citizens, it is not surprising that black people do not feel "safe" among white people in many social settings. Self-segregation is obviously one way to avoid racist assault and exploitation. There is so little documentation of the myriad ways racist habits of being make all black people potential targets of discrimination, petty exploitation, and in extreme cases racist assault in everyday life. Whether or not black folks feel better if they are exploited or wounded by another black person rather than someone of another race remains unclear. However, it is evident that racialized post-traumatic stress disorder often leads individual black people to experience tension, anxiety, and fear in the presence of white people, even those who are well meaning.

In her predominately black city, my younger sister G. has the positive experience of daily living among diverse classes of black folks yet she also has the negative experience of a culture of violence wherein criminal black folks prey on other black folks, often with impunity. Even though she knows from experience that she is not "safe" with someone solely because they are black (she and her loved ones have been threatened by black on black violence), she still feels more comfortable in segregated settings that are either all or predominately black. And she is not alone in this preference. Yet this way of thinking is the learned conservative pattern of identity politics which though useful during periods of extreme racist oppression as a basis of black solidarity and organized protest nowadays it actually undermines the struggle to end racism. Understanding this does not mean that as black people living within the political system of imperialist white supremacist capitalist patriarchy we do not have to be critically vigilant in social encounters with white people who have not unlearned racism but it does mean that we must simultaneously learn how to extend trust not only to white people, but to any non-black group.

White people do not all think the same about race; there are indeed individual white people who are as dedicated to ending racism as any anti-racist black person. Stereotyping all white folks, seeing every white person as a potential threat is as dehumanizing as judging all black folks by standards defined by negative racist stereotypes.

Of course it remains the responsibility of white citizens of this nation to work at unlearning and challenging the patterns of racist thought and behavior that are still a norm in our society. However, if whites and blacks alike do not remain mindful of the continual need to contest racist segregation and to work towards a racially integrated society free of white supremacy, then we will never live in beloved community. In the mid-sixties Lerone Bennett Jr. prophesied that there would come a time when citizens of this nation would have to decide between the American idea of democracy and fascism. He emphasized then that: "Real community is based on reciprocity of emotion and relation between individuals sharing a common vision of the possibilities and potentialities of man. The basic fact of race relations in America is that white people and Negroes do not belong to the same community." Currently in our nation Americans of all colors feel bereft of a sense of "belonging" to either a place or a community. Yet most people still long for community and that yearning is the place of possibility, the place where we might begin as a nation to think and dream anew about the building of beloved community.

This makes Berea, Kentucky, one of the best places to be because we do have community here and many of us are committed to striving for racial justice and an end to racial segregation. After living for several years in town, I decided to find myself a place in the hills, not far from town, a place for dreaming, writing, and retreating. When

I found the place, hilly land, lots of trees, a fancy cabin already built, a beautiful view of the lake that is our water source, I made ready to make my offer. But of course not before I inquired about the racial politics of the folks around me, the white folks living in trailers, who could possibly not welcome my presence. To begin with I met the "redneck" white hillbilly man who built this haven on the hill. At first meeting I had brought with me an older white woman friend, just in case E. was not friendly. We were a bit late coming up the hill to meet him but from the start it was acceptance at first sight. It seemed that the liberal white folks who had built the house were not certain that this hillbilly white man would want to work with a black woman so they had already let him know my color. He was not concerned.

I share the story of my meeting with E. because our working together, the friendship we have nurtured, the effort we have made to face our differences and resolve conflict, served as a catalyst for me to probe deeper notions of race and class, white supremacy, bonding across difference. It caused me to re-examine my past in relation to poor white hillbillies who were the first neighbors I knew growing up. We were taught back then that they hated black folks and we should stay away from them. No matter how nice or kind, we were taught that their appearance of kindness was really just a mask hiding their hateful intentions. No doubt there were many incidents of cruel racist assault of a black person by poor white people. And no doubt there were poor white people who lived in community with their black neighbors in the hills. The childhood socialization that taught us to simply fear and avoid contact with poor whites and not to use the same powers of discernment we might use with any group to decide whether they are a danger or not has left deep imprints. When they first heard that I had a cabin outside town, on a hill, my siblings first responded by interrogating me. They wanted to know if it was really safe for me to live among poor white folks in trailers. Understanding that I had expressed concerns about safety, about being a black woman alone in a place where most of the residents are poor and working-class white folks, I shared with my siblings that no one pondered whether I would feel safe in the predominately white neighborhood where I reside in town. The assumption being that privileged-class white folks are less likely to be racist than their poorer counterparts.

White supremacy is an active political system in our nation. It promotes and perpetuates racial discrimination and racist violence. Consequently, black people must remain aware and vigilant every day in our lives whether confronting people of color who have internalized racism or in all white settings where it is likely that many of those present have not unlearned racism. We must be discerning. At the same time we must not be paranoid or make blanket assumptions based on stereotypes about every white person we encounter. The reverse holds true for white people and their encounters with black people.

Those of us who truly believe racism can end, that white supremacist thought and action can be challenged and changed, understand that there is an element of risk as we work to build community across difference. The effort to build community in a social context of racial inequality (much of which is class based) requires an ethic of relational reciprocity, one that is anti-domination. With reciprocity all things do not need to be equal in order for acceptance and mutuality to thrive. If equality is evoked as the only standard by which it is deemed acceptable for people to meet across boundaries and create community, then there is little hope. Fortunately, mutuality is a more constructive and positive foundation for the building of ties that allow for differences in status, position, power, and privilege whether determined by race, class, sexuality, religion, or nationality.

Living in a community where many citizens work to end domination in all forms, including racial domination, a central aspect of our local culture is a willingness to be of service, especially to those who are for whatever reason among the disenfranchised. Dominator culture devalues the importance of service. Those of us who work to undo negative hierarchies of power understand the humanizing nature of service, understand that in the act of caregiving and caretaking we make ourselves vulnerable. And in that place of shared vulnerability there is the possibility of recognition, respect, and mutual partnership.

There are many locations in Berea where folks are doing the work of peace and justice, continuing the struggle to end domination in all its forms. We are fortunate to have a foundation in place for working anew to build beloved community. The ground has already been made ready here. Many seeds of John Fee's vision of ending racism and racial segregation, of creating beloved community where we live united by the understanding that "we are all people of one blood," lie dormant, simply waiting for growth to be nurtured. Hopefully Berea can one day be the same beacon light that it once was showing our nation what must take place if we are to create an anti-racist culture, if we are to live among beloved community.

8

Representations of Whiteness in the Black Imagination

Although there has never been any official body of black people in the United States who have gathered as anthropologists and/or ethnographers to study whiteness, black folks have, from slavery on, shared in conversations with one another "special" knowledge of whiteness gleaned from close scrutiny of white people. Deemed special because it was not a way of knowing that has been recorded fully in written material, its purpose was to help black folks cope and survive in a white supremacist society. For years, black domestic servants working in white homes, acting as informants, brought knowledge back to segregated communities — details, facts, observations, and psychoanalytic readings of the white Other.

Sharing the fascination with difference that white people have collectively expressed openly (and at times vulgarly) as they have traveled around the world in pursuit of the Other and Otherness, black people, especially those living during the historical period of racial apartheid and legal segregation, have similarly maintained steadfast and ongoing curiosity about the "ghosts," "the barbarians," these strange apparitions they were forced to serve. In the chapter on "Wildness" in *Shamanism, Colonialism, and the Wild Man*, Michael Taussig urges a stretching of our imagination and understanding of the Other to include inscriptions "on the edge of official history." Naming his critical project, identifying the passion he brings to the quest to know more deeply *you who are not ourselves*, Taussig explains:

> I am trying to reproduce a mode of perception — a way of seeing through a way of talking — figuring the world through dialogue that comes alive with sudden transforma-tive force in the crannies of everyday life's pauses and juxta-positions, as in the kitchens of the Putumayo or in the streets around the church in the Niña Maria. It is always a way of representing the world in the round-about "speech" of the college of things ... . It is a mode of perception that catches on the debris of history ... .

I, too, am in search of the debris of history. I am wiping the dust off past conversations to remember some of what was shared in the old days when black folks had little intimate contact with whites, when we were much more open about the way we connected white-ness with the mysterious, the strange, and the terrible. Of course, everything has changed. Now many black people live in the "bush of ghosts" and do not know themselves separate from whiteness. They do not know this thing we call "difference." Systems of domination, imperialism, colonialism, and racism actively coerce black folks to in-ternalize negative perceptions of blackness, to be self-hating. Many of us succumb to this. Yet blacks who imitate whites (adopting their val-ues, speech, habits of being, etc.) continue to regard whiteness with suspicion, fear, and even hatred. This contradictory longing to possess the reality of the Other, even though that reality is one that wounds and negates, is expressive of the desire to understand the mystery, to know intimately through imitation, as though such knowing worn like an amulet, a mask, will ward away the evil, the terror.

Searching the critical work of post-colonial critics, I found much writing that bespeaks the continued fascination with the way white minds, particularly the colonial imperialist traveler, perceive blackness, and very little expressed interest in representations of whiteness in the black imagination. Black cultural and social critics allude to such representations in their writing, yet only a few have dared to make explicit those perceptions of whiteness that they think will discomfort or antagonize readers. James Baldwin's collection of essays *Notes of a Native Son* explores these issues with a clarity and frankness that is no longer fashionable in a world where evocations of pluralism and diversity act to obscure differences arbitrarily imposed and maintained by white racist domination. Addressing the way in which whiteness exists without knowledge of blackness even as it collectively asserts control, Baldwin links issues of recognition to the practice of imperialist racial domination. Writing about being the first black person to visit a Swiss village with only white inhabitants in his essay "Stranger in the Village," Baldwin notes his response to the village's yearly ritual of painting individuals black who were then positioned as slaves and bought so that the villagers could celebrate their concern with converting the souls of the "natives":

> I thought of white men arriving for the first time in an African village, strangers there, as I am a stranger here, and tried to imagine the astounded populace touching their hair and marveling at the color of their skin. But there is a great difference between being the first white man to be seen by Africans and being the first black man to be seen by whites. The white man takes the astonishment as tribute, for he arrives to conquer and to convert the natives, whose inferiority in relation to himself is not even to be questioned, whereas I, without a thought of conquest, find myself among a people whose

culture controls me, has even, in a sense, created me, people
who have cost me more in anguish and rage than they will
ever know, who yet do not even know of my existence. The
astonishment with which I might have greeted them, should
they have stumbled into my African village a few hundred
years ago, might have rejoiced their hearts. But the astonishment with which they greet me today can only poison mine.

My thinking about representations of whiteness in the black
imagination has been stimulated by classroom discussions about
the way in which the absence of recognition is a strategy that fa-
cilitates making a group the Other. In these classrooms there have
been heated debates among students when white students respond
with disbelief, shock, and rage as they listen to black students talk
about whiteness, when they are compelled to hear observations,
stereotypes, etc., that are offered as "data" gleaned from close
scrutiny and study. Usually, white students respond with naive
amazement that black people critically assess white people from
a standpoint where "whiteness" is the privileged signifier. Their
amazement that black people watch white people with a critical
"ethnographic" gaze is itself an expression of racism. Often their
rage erupts because they believe that all ways of looking that highlight difference subvert the liberal belief in a universal subjectivity
(we are all just people) that they think will make racism disappear.
They have a deep emotional investment in the myth of "same-
ness," even as their actions reflect the primacy of whiteness as a
sign informing who they are and how they think. Many of them
are shocked that black people think critically about whiteness because racist thinking perpetuates the fantasy that the Other who is
subjugated, who is subhuman, lacks the ability to comprehend, to
understand, to see the working of the powerful. Even though the
majority of these students politically consider themselves liberals
and anti-racist, they too unwittingly invest in the sense of whiteness as mystery.

In white supremacist society, white people can "safely" imagine that they are invisible to black people since the power they have historically asserted, and even now collectively assert over black people, accorded them the right to control the black gaze. As fantastic as it may seem, racist white people find it easy to imagine that black people cannot see them if within their desire they do not want to be seen by the dark Other. One mark of oppression was that black folks were compelled to assume the mantle of invisibility, to erase all traces of their subjectivity during slavery and the long years of racial apartheid, so that they could be better, less threatening servants. An effective strategy of white supremacist terror and dehumanization during slavery centered around white control of the black gaze. Black slaves, and later manumitted servants, could be brutally punished for looking, for appearing to observe the whites they were serving, as only a subject can observe, or see. To be fully an object then was to lack the capacity to see or recognize reality. These looking relations were reinforced as whites cultivated the practice of denying the subjectivity of blacks (the better to dehumanize and oppress), of relegating them to the realm of the invisible. Growing up in a Kentucky household where black servants lived in the same dwelling with the white family who employed them, newspaper heiress Sallie Bingham recalls, in her autobiography *Passion and Prejudice*, "Blacks, I realized, were simply invisible to most white people, except as a pair of hands offering a drink on a silver tray." Reduced to the machinery of bodily physical labor, black people learned to appear before whites as though they were zombies, cultivating the habit of casting the gaze downward so as not to appear uppity. To look directly was an assertion of subjectivity, equality. Safety resided in the pretense of invisibility.

Even though legal racial apartheid no longer is a norm in the United States, the habits that uphold and maintain institutionalized white supremacy linger. Since most white people do not have to "see" black people (constantly appearing on billboards, television,

movies, in magazines, etc.) and they do not need to be ever on guard
nor to observe black people to be safe, they can live as though black
people are invisible, and they can imagine that they are also invisible
to blacks. Some white people may even imagine there is no representation of whiteness in the black imagination, especially one that
is based on concrete observation or mythic conjecture. They think
they are seen by black folks only as they want to appear. Ideolog-
ically, the rhetoric of white supremacy supplies a fantasy of whiteness. Described in Richard Dyer's essay "White," this fantasy makes
whiteness synonymous with goodness:

> Power in contemporary society habitually passes itself off as
> embodied in the normal as opposed to the superior. This is
> common to all forms of power, but it works in a peculiarly
> seductive way with whiteness, because of the way it seems
> rooted, in common-sense thought, in things other than eth-
> nic difference. ... Thus it is said (even in liberal textbooks)
> that there are inevitable associations of white with light and
> therefore safety, and black with dark and therefore danger,
> and that this explains racism (whereas one might well ar-
> gue about the safety of the cover of darkness, and the dan-
> ger of exposure to the light); again, and with more justice,
> people point to the Jewish and Christian use of white and
> black to symbolize good and evil, as carried still in such
> expressions as "a black mark," "white magic," "to blacken
> the character" and so on. Socialized to believe the fantasy,
> that whiteness represents goodness and all that is benign and
> non-threatening, many white people assume this is the way
> black people conceptualize whiteness. They do not imagine
> that the way whiteness makes its presence felt in black life,
> most often as terrorizing imposition, a power that wounds,
> hurts, tortures, is a reality that disrupts the fantasy of white-
> ness as representing goodness.

Collectively black people remain rather silent about representations of whiteness in the black imagination. As in the old days of
racial segregation where black folks learned to "wear the mask,"
many of us pretend to be comfortable in the face of whiteness only
to turn our backs and give expression to intense levels of discomfort. Especially talked about is the representation of whiteness as
terrorizing. Without evoking a simplistic essentialist "us and them"
dichotomy that suggests black folks merely invert stereotypical racist
interpretations so that black becomes synonymous with goodness
and white with evil, I want to focus on that representation of whiteness that is not formed in reaction to stereotypes but emerges as a response to the traumatic pain and anguish that remains a consequence
of white racist domination, a psychic state that informs and shapes
the way black folks "see" whiteness. Stereotypes black folks maintain about white folks are not the only representations of whiteness
in the black imagination. They emerge primarily as responses to
white stereotypes of blackness. Lorraine Hansberry argues that black
stereotypes of whites emerge as a trickle-down process of white stereotypes of blackness, where there is the projection onto an Other
all that we deny about ourselves. In *Young, Gifted, and Black*, she
identifies particular stereotypes about white people that are com-
monly cited in black communities and urges us not to "celebrate this
madness in any direction":

> Is it not "known" in the ghetto that white people, as an
> entity, are "dirty" (especially white women — who never
> seem to do their own cleaning); inherently "cruel" (the
> cold, fierce roots of Europe; who else could put all those
> people into ovens *scientifically*); "smart" (you really have to
> hand it to the m.f.'s), and anything *but* cold and passionless
> (because look who has had to live with little else than their
> passions in the guise of love and hatred all these centuries)?
> And so on.

Stereotypes, however inaccurate, are one form of representation. Like fictions, they are created to serve as substitutions, standing in for what is real. They are there not to tell it like it is but to invite and encourage pretense. They are a fantasy, a projection onto the Other that makes them less threatening. Stereotypes abound when there is distance. They are an invention, a pretense that one knows when the steps that would make real knowing possible cannot be taken or are not allowed.

Looking past stereotypes to consider various representations of whiteness in the black imagination, I appeal to memory, to my earliest recollections of ways these issues were raised in black life. Returning to memories of growing up in the social circumstances created by racial apartheid, to all black spaces on the edges of town, I reinhabit a location where black folks associated whiteness with the terrible, the terrifying, the terrorizing. White people were regarded as terrorists, especially those who dared to enter that segregated space of blackness. As a child, I did not know any white people. They were strangers, rarely seen in our neighborhoods. The "official" white men who came across the tracks were there to sell products, Bibles, and insurance. They terrorized by economic exploitation. What did I see in the gazes of those white men who crossed our thresholds that made me afraid, that made black children unable to speak? Did they understand at all how strange their whiteness appeared in our living rooms, how threatening? Did they journey across the tracks with the same "adventurous" spirit that other white men carried to Africa, Asia, to those mysterious places they would one day call the "third world"? Did they come to our houses to meet the Other face to face and enact the colonizer role, dominating us on our own turf?

Their presence terrified me. Whatever their mission, they looked too much like the unofficial white men who came to enact rituals of terror and torture. As a child, I did not know how to tell them apart, how to ask the "real white people to please stand up." The terror that I felt is one black people have shared. Whites learn about it secondhand. Confessing in *Soul Sister* that she too began to feel this terror after changing her skin to appear "black" and going to live in the South, Grace Halsell described her altered sense of whiteness:

> Caught in this climate of hate, I am totally terrorstricken, and I search my mind to know why I am fearful of my own people. Yet they no longer seem my people, but rather the "enemy" arrayed in large numbers against me in some hostile territory.... My wild heartbeat is a secondhand kind of terror. I know that I cannot possibly experience what *they*, the black people, experience....

Black folks raised in the north do not escape this sense of terror. In her autobiography, *Every Good-bye Ain't Gone*, Itabari Njeri begins the narrative of her northern childhood with a memory of southern roots. Traveling south as an adult to investigate the murder of her grandfather by white youth who were drag racing and ran him down in the streets, Njeri recalls that for many years "the distant and accidental violence that took my grandfather's life could not compete with the psychological terror that had begun to engulf my own." Ultimately, she begins to link that terror with the history of black people in the United States, seeing it as an imprint carried from the past to the present:

> As I grew older, my grandfather assumed mythic proportions in my imagination. Even in absence, he filled my room like music and watched over me when I was fearful. His fantasized presence diverted thoughts of my father's drunken rages. With age, my fantasizing ceased, the image of my grandfather faded. What lingered was the memory of his caress, the pain of something missing in my life, wrenched away by reckless white youths. I had a growing sense — the beginning of an inevitable comprehension — that this society deals blacks a disproportionate share of pain and denial.

Njeri's journey takes her through the pain and terror of the past, only the memories do not fade. They linger as does the pain and bitterness: "Against a backdrop of personal loss, against the evidence of history that fills me with a knowledge of the hateful behavior of whites toward blacks, I see the people of Bainbridge. And I cannot trust them. I cannot absolve them." If it is possible to conquer terror through ritual reenactment, that is what Njeri does. She goes back to the scene of the crime, dares to face the enemy. It is this confrontation that forces the terror of history to loosen its grip.

To name that whiteness in the black imagination is often a representation of terror. One must face written histories that erase and deny, that reinvent the past to make the present vision of racial harmony and pluralism more plausible. To bear the burden of memory one must willingly journey to places long uninhabited, searching the debris of history for traces of the unforgettable, all knowledge of which has been suppressed. Njeri laments that "nobody really knows us." She writes, "So institutionalized is the ignorance of our history, our culture, our everyday existence that, often, we do not even know ourselves." Theorizing black experience, we seek to uncover, restore, as well as to deconstruct, so that new paths, different journeys, are possible. Indeed, Edward Said, in his essay "Traveling Theory," argues that theory can "threaten reification, as well as the entire bourgeois system on which reification depends, with destruction." The call to theorize black experience is constantly challenged and subverted by conservative voices reluctant to move from fixed locations. Said reminds us:

> Theory ... is won as the result of a process that begins when consciousness first experiences its own terrible ossification in the general reification of all things under capitalism; then when consciousness generalizes (or classes) itself as something opposed to other objects, and feels itself as contradiction to (or crisis within) objectification, there emerges a consciousness of change in the *status quo*; finally, moving

toward freedom and fulfillment, consciousness looks ahead to complete selfrealization, which is of course the revolutionary process stretching forward in time, perceivable now only as theory or projection.

Traveling, moving into the past, Njeri pieces together fragments. Who does she see staring into the face of a southern white man who was said to be the murderer? Does the terror in his face mirror the look of the unsuspecting black man whose death history does not name or record? Baldwin wrote that "people are trapped in history and history is trapped in them." There is then only the fantasy of escape, or the promise that what is lost will be found, rediscovered, and returned. For black folks, reconstructing an archaeology of memory makes return possible, the journey to a place we can never call home even as we reinhabit it to make sense of present locations. Such journeying cannot be fully encompassed by conventional notions of travel.

Spinning off from Said's essay, James Clifford, in "Notes on Travel and Theory," celebrates the idea of journeying, asserting:

> This sense of worldly, "mapped" movement is also why it may be worth holding on to the term "travel," despite its connotations of middle class "literary" or recreational journeying, spatial practices long associated with male experiences and virtues. "Travel" suggests, at least, profane activity, following public routes and beaten tracks. How do different populations, classes and genders travel? What kinds of knowledges, stories, and theories do they produce? A crucial research agenda opens up.

Reading this piece and listening to Clifford talk about theory and travel, I appreciated his efforts to expand the travel/theoretical frontier so that it might be more inclusive, even as I considered that to answer the questions he poses is to propose a deconstruction of the conventional sense of travel, and put alongside it, or in its place, a theory of the journey that would expose the extent to which holding on to the concept of "travel" as we know it is also a way to hold on to imperialism.

For some individuals, clinging to the conventional sense of travel allows them to remain fascinated with imperialism, to write about it, seductively evoking what Renato Rosaldo aptly calls, in *Culture and Truth*, "imperialist nostalgia." Significantly, he reminds readers that "even politically progressive North American audiences have enjoyed the elegance of manners governing relations of dominance and subordination between the 'races.'" Theories of travel produced outside conventional borders might want the Journey to become the rubric within which travel, as a starting point for discourse, is associated with different headings — rites of passage, immigration, enforced migration, relocation, enslavement, and homelessness. "Travel" is not a word that can be easily evoked to talk about the Middle Passage, the Trail of Tears, the landing of Chinese immigrants, the forced relocation of Japanese Americans, or the plight of the homeless. Theorizing diverse journeying is crucial to our understanding of any politics of location. As Clifford asserts at the end of his essay:

> Theory is always written from some "where," and that "where" is less a place than itineraries: different, concrete histories of dwelling, immigration, exile, migration. These include the migration of third world intellectuals into the metropolitan universities, to pass through or to remain, changed by their travel but marked by places of origin, by peculiar allegiances and alienations.

Listening to Clifford "playfully" evoke a sense of travel, I felt such an evocation would always make it difficult for there to be recognition of an experience of travel that is not about play but is an encounter with terrorism. And it is crucial that we recognize that the hegemony of one experience of travel can make it impossible to articulate another experience or for it to be heard. From certain standpoints, to travel is to encounter the terrorizing force of white supremacy. To tell my "travel" stories, I must name the movement from racially segregated southern community, from rural black Baptist origin, to prestigious white university settings. I must be able to speak about what it is like to be leaving Italy after I have given a talk on racism and feminism, hosted by the parliament, only to stand for hours while I am interrogated by white officials who do not have to respond when I inquire as to why the questions they ask me are different from those asked the white people in line before me. Thinking only that I must endure this public questioning, the stares of those around me, because my skin is black, I am startled when I am asked if I speak Arabic, when I am told that women like me receive presents from men without knowing what those presents are. Reminded of another time when I was strip-searched by French officials, who were stopping black people to make sure we were not illegal immigrants and/or terrorists, I think that one fantasy of whiteness is that the threatening Other is always a terrorist. This projection enables many white people to imagine there is no representation of whiteness as terror, as terrorizing. Yet it is this representation of whiteness in the black imagination, first learned in the narrow confines of poor black rural community, that is sustained by my travels to many different locations.

To travel, I must always move through fear, confront terror. It helps to be able to link this individual experience to the collective journeying of black people, to the Middle Passage, to the mass migration of southern black folks to northern cities in the early part of the twentieth century. Michel Foucault posits memory as a site of resistance. As Jonathan Arac puts it in his introduction to *Postmodernism and Politics*, the process of remembering can be a practice which "transforms history from a judgement on the past in the name of a present truth to a 'counter-memory' that combats our current modes of truth and justice, helping us to understand and change the present by placing it in a new relation to the past." It is useful, when theorizing black experience, to examine the way the concept of "terror" is linked to representations of whiteness.

In the absence of the reality of whiteness, I learned as a child that to be "safe" it was important to recognize the power of whiteness, even to fear it, and to avoid encounter. There was nothing terrifying about the sharing of this knowledge as survival strategy; the terror was made real only when I journeyed from the black side of town to a predominantly white area near my grandmother's house. I had to pass through this area to reach her place. Describing these journeys "across town" in the essay "Homeplace: A Site of Resistance," I remembered:

> It was a movement away from the segregated blackness of our community into a poor white neighborhood. I remember the fear, being scared to walk to Baba's, our grandmother's house, because we would have to pass that terrifying whiteness — those white faces on the porches staring us down with hate. Even when empty or vacant those porches seemed to say *danger*, you do not belong here, you are not safe.

Oh! that feeling of safety, of arrival, of homecoming when we finally reached the edges of her yard, when we could see the soot black face of our grandfather, Daddy Gus, sitting in his chair on the porch, smell his cigar, and rest on his lap. Such a contrast, that feeling of arrival, of homecoming — this sweetness and the bitterness of that journey, that constant reminder of white power and control. Even though it was a long time ago that I made this journey, associations of whiteness with terror and the terrorizing remain. Even though I live and move in spaces where I am surrounded by whiteness, there is no comfort that makes the terrorism disappear. All black people in the United States, irrespective of their class status or politics, live with the possibility that they will be terrorized by whiteness.

This terror is most vividly described by black authors in fiction writing, particularly Toni Morrison's novel *Beloved*. Baby Suggs, the black prophet, who is most vocal about representations of whiteness, dies because she suffers an absence of color. Surrounded by a lack, an empty space, taken over by whiteness, she remembers: "Those white things have taken all I had or dreamed and broke my heartstrings too. There is no bad luck in the world but white folks." If the mask of whiteness, the pretense, represents it as always benign, benevolent, then what this representation obscures is the representation of danger, the sense of threat. During the period of racial apartheid, still known by many folks as Jim Crow, it was more difficult for black people to internalize this pretense, hard for us not to know that the shapes under white sheets had a mission to threaten, to terrorize. That representation of whiteness, and its association with innocence, which engulfed and murdered Emmett Till, was a sign; it was meant to torture with the reminder of possible future terror. In Morrison's *Beloved*, the memory of terror is so deeply inscribed on the body of Sethe and in her consciousness, and the association of terror with whiteness is so intense, that she kills her young so that they will never know the terror. Explaining her actions to Paul D., she tells him that it is her job "to keep them away from what I know is terrible." Of course Sethe's attempt to end the historical anguish of black people only reproduces it in a different form. She conquers the terror through perverse reenactment, through resistance, using violence as a means of fleeing from a history that is a burden too great to bear.

It is the telling of our history that enables political self-recovery. In contemporary society, white and black people alike believe that racism no longer exists. This erasure, however mythic, diffuses the representation of whiteness as terror in the black imagination. It allows for assimilation and forgetfulness. The eagerness with which contemporary society does away with racism, replacing this recognition with evocations of pluralism and diversity that further mask reality, is a response to the terror. It has also become a way to perpetuate the terror by providing a cover, a hiding place. Black people still feel the terror, still associate it with whiteness, but are rarely able to articulate the varied ways we are terrorized because it is easy to silence by accusations of reverse racism or by suggesting that black folks who talk about the ways we are terrorized by whites are merely evoking victimization to demand special treatment.

When I attended a recent conference on cultural studies, I was reminded of the way in which the discourse of race is increasingly divorced from any recognition of the politics of racism. Attending the conference because I was confident that I would be in the company of like-minded, "aware," progressive intellectuals, I was disturbed when the usual arrangements of white supremacist hierarchy were mirrored both in terms of who was speaking, of how bodies were arranged on the stage, of who was in the audience. All of this revealed the underlying assumptions of what voices were deemed worthy to speak and be heard. As the conference progressed, I began to feel afraid. If these progressive people, most of whom were white, could so blindly reproduce a version of the status quo and not "see" it, the thought of how racial politics would be played out "outside" this arena was horrifying. That feeling of terror that I had known so intimately in my childhood surfaced. Without even considering whether the audience was able to shift from the prevailing standpoint and hear another perspective, I talked openly about that sense of terror. Later, I heard stories of white women joking about how ludicrous it was for me (in their eyes I suppose I represent the "bad" tough black woman) to say I felt terrorized. Their inability to conceive that my terror, like that of Sethe's, is a response to the legacy of white domination and the contemporary expressions of white supremacy is an indication of how little this culture really understands the profound psychological impact of white racist domination.

At this same conference, I bonded with a progressive black woman and her companion, a white man. Like me, they were troubled by the extent to which folks chose to ignore the way white supremacy was informing the structure of the conference. Talking with the black woman, I asked her: "What do you do, when you are tired of confronting white racism, tired of the day-to-day incidental acts of racial terrorism? I mean, how do you deal with coming home to a white person?" Laughing she said, "Oh, you mean when I am suffering from White People Fatigue Syndrome? He gets that more than I do." After we finish our laughter, we talk about the way white people who shift locations, as her companion has done, begin to see the world differently. Understanding how racism works, he can see the way in which whiteness acts to terrorize without seeing himself as bad, or all white people as bad, and all black people as good. Repudiating us-and-them dichotomies does not mean that we should never speak of the ways observing the world from the standpoint of "whiteness" may indeed distort perception, impede understanding of the way racism works both in the larger world as well as in the world of our intimate interactions.

In *The Post-Colonial Critic*, Gayatri Spivak calls for a shift in locations, clarifying the radical possibilities that surface when positionality is problematized. She explains that "what we are asking for is that the hegemonic discourses, and the holders of hegemonic discourse, should dehegemonize their position and themselves learn how to occupy the subject position of the other." Generally, this process of repositioning has the power to deconstruct practices of racism and make possible the disassociation of whiteness with terror in the black imagination. As critical intervention it allows for the recognition that progressive white people who are anti-racist might be able to understand the way in which their cultural practice reinscribes white supremacy without promoting paralyzing guilt or denial. Without the capacity to inspire terror, whiteness no longer signifies the right to dominate. It truly becomes a benevolent absence. Baldwin ends his essay "Stranger in the Village" with the declaration: "This world is white no longer, and it will never be white again." Critically examining the association of whiteness as terror in the black imagination, deconstructing it, we both name racism's impact and help to break its hold. We decolonize our minds and our imaginations.

9

Drive through Tobacco

Riding in the car, away from the town, riding in the country, we were surrounded by fields and fields of tobacco. Growing up in Kentucky I learned the reverence for the tobacco plant that had been handed down from generation to generation. In those days tobacco was not demonized. Tobacco was a sacred plant, cherished and deemed precious by the old folks who knew its properties and its potentialities.

I cannot recall any time in my childhood when tobacco did not have meaning and presence. Whether it came from watching Big Mama smoke her pipe, or emptying the coffee cans that were used to spit out chewing tobacco, or watching Mama's mother Baba braid tobacco leaves for use to ward off bugs, or watching Aunt Margaret hand-roll tobacco for cigarettes and cigars, the odor of tobacco permeated our lives and touches me always with the scent of memory. Once upon a time tobacco ruled the economy of many small Kentucky towns. For poor illiterate black folks picking tobacco was down and dirty work but it let one bring home ready cash, extra money. The history of black folks and the history of tobacco like braided leaves were once deeply intertwined. And even though that history, like so many aspects of our ancestral past is unsung, buried, the old folks remember, think of the past, and smell the fragrance of tobacco.

Tobacco planting, harvesting, picking, and curing played a major role in the drama of enslaved Africans throughout the southern states. While displaced Africans encountered many odd ways and strange appetites in the so called "new world," tobacco was a "familiar." When tobacco first came to Africa, its power and pleasure quickly spread throughout the continent. At the very onset Africans conferred on the plant mystical and magical powers. Both in Africa and parts of South America, tobacco shamans used the plant ceremoniously and ritualistically. Used to heal, to bless, to protect, tobacco had divine status. In his lengthy work *Tobacco: A Cultural History of How An Exotic Plant Seduced Civilization*, cultural critic Iain Gately informs readers that: "Tobacco played a central role in the spiritual training of shamans ... A tobacco shaman used the weed in every aspect of his art." In North America the indigenous Native Indians also considered tobacco a sacred plant and many tribal groups continue to honor the spiritual nature of tobacco both as powerful legacy and potent element of current traditions. Writing about the history of tobacco in this country, Gately emphasizes that it was "a defining habit of the diverse tribes and civilizations that occupied pre-Columbian North American" as "everyone of its cultures living and vanished, used tobacco." Indians used tobacco medicinally, to clear the skin, to cleanse the body, to purge. While tobacco was more often than not the sole purview of men in cultures around the world, on the African continent men and women used tobacco with equal relish and fervor, seeking empowerment and solace from holy smoke.

Spiritually, people of color globally have viewed and continue to view tobacco as a way to be initiated into the spirit world. Many shamans both past and present use tobacco to trance or even go toward death, then resurrect as a sign of their powers. Exploring the link between tobacco and religion, Gately informs readers that: "Ritual smoke blowing, by which a shaman might bestow a blessing or protection against enemies was intended to symbolize a transformation in which the tobacco smoke represented a guiding spirit, and this is reminiscent of Christian ritual, whereby wine and bread are transubstantiated by a priest into body and blood of Christ himself." In Native American culture smoke signals are a nonlocal means of cosmic communication. In *Reinventing Medicine*, Larry Dossey explains that "the function of the smoke signal was only to get everyone's attention so that distant, mind-to-mind communications might then take place." According to Dossey, "the possibility that the mind might function at a distance, outside the confines of the brain and body and not just in dreams, is taken for granted in most of what we call 'native' cultures." Tobacco and tobacco smoke bring the promise of transcending one's limitations. Hence Native culture's reverence and respect for tobacco.

At the onset of the powerful memoir *Black Elk Speaks: Being The Life Story Of A Holy Man Of The Oglala Sioux*, this wise tribal elder tells the story of how sacred visions come to him as he smokes the "pipe with a bison calf carved on one side to mean the earth that bears and feeds us, and with twelve eagle feathers hanging from the stem to mean the sky and the twelve moons, and these were tied with a grass that never breaks." When he smokes the pipe after offering "it to the powers that are one Power, and sending forth a voice to them, we shall smoke together." He talks with the Great Spirit declaring: "Great Spirit, Great Spirit, my Grandfather, all over the earth the faces of living things are all alike. With tenderness have these come up out of the ground. Look upon these faces of children without number and with children in their arms, that they may face the winds and walk the good road to the day of quiet." After his prayers to the Great Spirit, he welcomes tribal companions telling them, "let us smoke together so that there may be only good between us."

Within popular culture in the United States negative media representations of Native Americans, especially television images, have changed little. Smoking a peace pipe is consistently caricatured. Clearly, dominator culture mocked the Native belief in oneness with nature and the naturalness of peace when slaughtering Indians was deemed justified because they were supposedly a savage and violent people. Yet surviving artifacts relating to tobacco (beautifully carved pipes) attest to the spiritual significance of smoking.

Though little is written of the role tobacco played as social currency between enslaved Africans and white slavers, demands for workers to plant and harvest the crop was so intense in the eighteenth and nineteenth centuries that more slaves were purchased to do the dirty work of planting and harvesting tobacco and thereby increasing an owner's wealth. Gately reminds readers that "tobacco's importance was not limited to the colonies of the south" as it soon became a principal crop for export. Indeed, he contends "the weed had given the colonies a place in the world." Tobacco — smoked, chewed, or sniffed was a source of tremendous power to a growing capitalist culture of greed in the United States. For the slaves who would work to the bone to make this crop plentiful, whose existence was fiercely nomadic as planters moved to find fresh soil, the one sure reward of harsh dirty labor was the freedom to use a bit of tobacco that, like gold, was precious and hard come by.

Among enslaved black folks, men and women often smoked their tobacco in pipes. No doubt smoking one's pipe at the end of a grueling work day, sitting quietly in a meditative pose, offered enslaved Africans a way to psychically leave their concrete harsh circumstances and literally be somewhere else. The reverence for tobacco as a sacred plant that had been a central part of the African experience was sustained by both the small numbers of African explorers who came to the "new world" before Columbus and the newly enslaved Africans. Even though their work on tobacco farms was harsh and life threatening, enslaved black folk were still able to retain the ceremonial culture of tobacco that was a distinctive feature of life before exile.

These cultural retentions were carried on in black life after slavery ended. Among the elders in my family the tobacco plant had pride of place. Tobacco could be used to disinfect, to serve as a deterrent for bugs. Often braided leaves of tobacco would be placed in trunks of clothing and other linens to keep dust mites and fabric-eating bugs

from destroying costly cloth. Tobacco was useful and it was a source
of pleasure. Even though I was not seduced by the world of smoking
or chewing tobacco growing, I was seduced both by the beauty of
tobacco growing, curing, hanging, or braided. In that world where
women smoked, chewed, and dipped as much as men, nothing pleased
me more than to be allowed to tenderly handle precious tobacco leaves
and put them in Big Mama's pipe just so, making sure not to waste. In
those days smoking was not viewed as the health hazard we know it
to be today. However, it must be stated that the southern black folks
who harvested and cured tobacco plants with no harmful pesticides or
additives used tobacco and usually lived long lives. For them, tobacco
was deeply healing. Danger from smoking came into their lives when
they begin to smoke packaged tobacco and store-bought cigarettes
and cigars, when they became addicted to smoking. These old black
folks worked and exercised; they would never have embraced a lifestyle where they simply indulged in the pleasures of smoking if that
pleasure did not come at the end of a hard day of physical labor.

Nostalgia for my Kentucky childhood often focuses my mind on
two distinct memories — the world of tobacco and the world of quiltmaking. Both are associated in my mind with simple living and simple abundance. Both are associated with comfort of mind, body, and
soul. Big Mama, Daddy's grandmother, who loved us uncondition-
ally, was short and squat. She came wearing a perfectly ironed apron,
with pockets wide enough for tobacco and pipe. Burley tobacco grew
all around us. The aroma of her pipe tobacco, the clouds of smoke and
most importantly the contentment surrounding her body and being
would have made it difficult for the grandchildren to see tobacco as
threatening. It was part of her personal magic and majesty.

Perhaps had I not been a sickly child constantly faced with
breathing problems, smoking tobacco would have held greater al-
lure. However, I was seduced by the aesthetics of tobacco and by
its presence in our lives and in our African and Native American
origins as a sacred plant. For many people of color who claim the old
sacred ways of our elders, who work to restore meaningful traditions

ravaged by white imperialism and colonization, it has been essential
to hold to our ancient understandings of the mystical and magical
power of tobacco. This has not been an easy task in a world where
tobacco laced with poisons addicts, destroys, and kills, where abusive
use of tobacco leads to disease and death.

The system of imperialist white supremacist capitalist patriarchy
that took the tobacco plant and made it into a product to be marketed
solely for excess profit is continually critiqued, perhaps more so than
any other plant drug because it is legal. And because the tobacco
industry, through seductive marketing and advertising, invites consumers to choose death. Ironically, capitalist marketing seduces consumers using the same subliminal suggestions which convey to the
public that tobacco has mystical and magical powers that were used
more overtly in ancient times to seduce Native people everywhere in
the world. However, this contemporary marketing completely severs
tobacco from its roots as a healing and sacred plant. Just as the colonization of Native and African peoples required that they be stripped
of their language, identity, and dehumanized, the tobacco plant un-
derwent a similar process. In *Food of the Gods*, Terence McKenna
describes the way the world of tobacco changed with the coming of
the white man, the colonizer: "Tobacco was the first and most immediate payoff of the discovery of the New World. On November 2,
1492, less than a month after his first arrival in the New World, Columbus landed on the north coast of Cuba ... Scouts returned with
an account of men and women who partially inserted burning rolls
of leaves into their nostrils. These burning rolls were called tobaccos
and consisted of dry herbs wrapped up in a large dry leaf. They were
lit at one end, and the people sucked at the other and 'drank the
smoke,' or inhaled something that was utterly unknown in Europe."

Tobacco as a capitalist industry has been subjected to all the
machinations of imperialist corruption and greed. Without a doubt,
tobacco is the most widely consumed plant drug on earth. Removed
from all its medicinal legacy tobacco has come to be demonized solely
as a product that kills. And wrongly used, addictively used, it will

indeed take users on a path that will lead to disease and ultimately death. If current smoking trends continue globally, in this century alone one billion people will die from tobacco-related diseases. According to the World Health Organization, India and China now account for forty percent of the world's smoking population. In the United States there has been tremendous consciousness raising about the dangers of tobacco and the corruption of the tobacco industry, but there has been little or no effort to separate the tobacco plant and its positive attributes from all that is negative. In the recent Al Gore documentary, *An Inconvenient Truth*, a film about global warming, he shared the heartrending story of his sister's death from lung cancer. She smoked at an early age. The film showed his grieving father, a long-time tobacco grower, facing the hard truth about the role tobacco played in causing the death of his daughter. And viewers hear commentary about his father's decision to stop tobacco farming. There is no effort in the film to depict an alternative view of tobacco, one that would separate the plant from the tobacco industry and its calculated use of poisonous additives or from addictive smoking.

The Native people in our culture who continue to regard the tobacco plant reverently have no public voice. And the tyranny of fundamentalist Christianity usually obscures the presence of sacred traditions that are not Bible based. None of the anti-tobacco pundits offer the possibility that were young folk, who are especially vulnerable to advertising and marketing that targets their desires, to be taught alternative ways to think and dream about tobacco, ways that would teach respect for this wondrous yet potentially dangerous plant, they would not become addicted. The public could be taught to relate tobacco use to sacred traditions. If this were the new culture of tobacco, the public would have the opportunity to choose what their relationship to smoke and smoking might be rather than to be led mindlessly into the culture of death created by mainstream poisoned tobacco. When imperialist tobacco producers received public sanction to go to underdeveloped countries to grow and harvest their crop, without the health regulations regarding insecticides and poisonous addicting additives that "might" be imposed in the United States, it signaled the end of major tobacco growing in this country. Tobacco, the crop that had once called the world's attention to the United States' market, has little meaning in cultural iconography today. My home state of Kentucky once produced huge quantities of burley tobacco, bringing huge revenue into its coffers, money that primarily made the rich richer, but that day has long passed. Tobacco is no longer vital to the economy of Kentucky. The miles and miles of farmland where tobacco grew, seemingly on and on into eternity, that were visible during my childhood, are no longer. Once upon a time one could walk into any small town Kentucky store and find tobacco hanging in all its glory, beautifully braided leaves to be shared as gesture of plenty and regard; nowadays tobacco has no place. Certainly this beautiful plant cannot line the walls of superstores or drive-through tobacco marts.

Distinguishing, as I do, the harmful effects of smoking, of addictive use, from the tobacco plant, I mourn the loss of those tobacco fields for all that they stood for in our childhood. First and foremost, they represented the bounty of nature, the richness which the earth offers to us. And so, as the elders taught, we are given the beauty of smoke, the aroma of tobacco, to enhance life. When I talk with my siblings, those who worked harvesting and stripping tobacco when we were young, they remember the dust, the ache in their bones from bending, the cold air on the loosening floor. But they recall as well the culture of tobacco that freely gave us images of beauty, rows of tobacco leaves hanging in a barn, green fields and our young childish voices wanting to hear the plant growing everywhere identified — tobacco. Finding words to express the aesthetics of the tobacco plant, the beauty and bounty of tobacco leaves hanging in barns, is no simple matter in today's world where tobacco is mainly viewed with disrespect and disregard, if recognized and remembered at all. In my book of love poems, *When Angels Speak of Love*, tobacco is muse. In my imagining I dream of "braided tobacco leaves twisted hung time on the loosening floor, time stripping, time drying, time turning, sheets of brown, time turning away, and all the time love, the smell of smoke between us." Whether through simple nostalgia or meaningful cultural memory, the tobacco plant is worthy to be cherished.

Globally, from its origins to the present day, tobacco and tobacco use have been linked to freedom. Like the white Kentucky abolitionist politician Henry Clay, who neither, as Gately put it, "slavered nor spat," I who will never smoke, dip, or chew understand intimately the lure of tobacco. A powerful advocate of universal human rights, Clay visited Cuba and was so welcomed that in 1850 a cigar brand was named after him as a gesture of respect. When James Weldon Johnson's novel *The Autobiography of An Ex-Coloured Man* was published in 1912, he included a fictive portrait of cigar making. His main character remembers that: "At first the heavy odour of tobacco sickened me, but when I became accustomed to it, I liked the smell ... ." In the early nineteenth century, black men found they could make ready cash working in tobacco more than in other trades where they were cruelly discriminated against. Johnson's protagonist associates tobacco with freedom, commenting: "Cigar-making was a rather independent trade; the men went to work when they pleased and knocked off when they felt like doing so." Southern black males working in tobacco fields and tobacco plants were able to make a better income than they could make doing other forms of service labor. Even though dominator culture made tobacco use in the Western world a patriarchal male privilege, using tobacco has been for females of all ages a way to assert independence. How life enhancing it would be if tobacco were used by females only in coming of age rituals. Sadly, as equal consumers of tobacco, females court death and disease. Because nicotine is so highly addictive, it is only the fortunate who can use tobacco non-addictively.

Clearly, all over the world people on planet earth use tobacco. It is the most democratic of all plant drugs. It is the drug that it is

legal to use and easy to obttain. McKenna argues that a more strin-
gent tax on tobacco would limit use. In the final pages of *Food of the
Gods* he reminds readers that folks will always seek ecstasy (to stand
outside) through the use of mood-altering drugs. And he reminds
us that this longing is indeed a yearning that is religious in nature.
He contends: "Help from nature means recognizing that the satis-
faction of the religious impulse comes not from ritual, and still less
from dogma, but rather, from a fundamental kind of experience —
the experience of symbiosis with hallucinogenic plants, and through
them, symbiosis with the whole of planetary life ...Without the escape hatch into the transcendental and transpersonal realm that is
provided by plant-based hallucinogens, the human future would be
bleak indeed." McKenna suggests that survival lies not in forbidding
tobacco but in creating a context for meaningful use of plant drugs,
uses that could aid in restoring a non-dominating relationship to
nature. Sharing this insight he explains: "The shamanic plants and
the worlds that they reveal are the worlds from which we imagine
that we came long ago, worlds of light and power and beauty ... We
can claim this prodigal legacy only as quickly as we can remake our
language and ourselves. Remaking out language means rejecting the
image of ourselves inherited from dominator culture ... Nature is
not our enemy, to be raped and conquered. Nature is ourselves, to
be cherished and explored." His words echo the teachings about our
right relationship to nature that I received from Kentucky elders.

In my life in Kentucky today, I know no tobacco farmers. Journeying down country roads I come upon lone tobacco barns loaded
with treasure and I feel again the sense of wonder and awe that this
union of human and plant produces. It is for a moment of interbeing
whose beauty blesses me. And I do not want to forget. I want to hold
forever in my hand braided tobacco, planted by my elders, braided
by the beloved hands of Baba, Mama's mother. I want to cherish the
tobacco plant — let its sacred appeal be the legacy that calls to me.

10

Earthbound: On Solid Ground

Kentucky hills were the place of my early childhood. Surrounded by a wilderness of honeysuckle, wild asparagus, and sheltering trees, bushes shielding growing crops, the huge garden of a black landowner. Our concrete house on the hill, a leftover legacy from oil drilling, from the efforts of men to make the earth yield greater and greater profit stood as a citadel to capitalism's need for a new frontier. A child of the hills, I was taught early on in my life the power in nature. I was taught by farmers that wilderness land, the untamed environment, can give life and it can take life. In my girlhood I learned to watch for snakes, wildcats roaming, plants that irritate and poison. I know instinctively; I know because I am told by all knowing grown-ups that it is humankind and not nature that is the stranger on these grounds. Humility in relationship to nature's power made survival possible.

Coming from "backwoods" folks, Appalachian outlaws, as a child I was taught to understand that those among us who lived organically, in harmony and union with nature were marked with a sensibility that was distinct, and downright dangerous. Backwoods folks tend to ignore the rules of society, the rules of law. In the backwoods one learned to trust only the spirit, to follow where the spirit moved. Ultimately, no matter what was said or done, the spirit called to us from a place beyond words, from a place beyond man-made law. The wild spirit of unspoiled nature worked its way in to the folk of the backwoods, an ancestral legacy, handed down from generation to generation. And its fundamental gift the cherishing of that which is most precious, freedom. And to be fully free one had to embrace the organic rights of the earth.

Humankind, no matter how powerful, cannot take away the rights of the earth. Ultimately, nature rules. That is the great demo-cratic gift earth offers us — that sweet death to which we all inevi-tably go — into that final communion. No race, no class, no gender, nothing can keep any of us from dying into that death where we are made one. To tend the earth is always then to tend our destiny, our freedom, and our hope.

These lessons of my girlhood were the oppositional narratives that taught me to care for the earth, to respect country folk. This respect for the earth, for the country girl within, stood me in good stead when I left this environment and entered a world beyond the coun-try town I was raised in. It was only when I left home, that country place where nature's splendors were abundant and not yet destroyed, that I understood for the first time the contempt for country folk that abounds in our nation. That contempt has led to the cultural disrespect for the farmer, for those who live simply in harmony with nature. Writer, sometime farmer, and poet Wendell Berry, another Kentuckian, who loves our land, writes in *Another Turn of the Crank* in the essay "Conserving Communities" that "Communists and capitalists are alike in their contempt for country people, country life, and country places."

Before the mass migrations to northern cities in the early nine-teen hundreds, more than ninety percent of all black folks lived in the agrarian South. We were indeed a people of the earth. Working the land was the hope of survival. Even when that land was owned by white oppressors, master and mistress, it was the earth itself that protected exploited black folks from dehumanization. My sharecropping granddaddy Jerry would walk through neat rows of crops and tell me, "I'll tell you a secret, little girl. No man can make the sun or the rains come — we can all testify. We can all see that ultimately we all bow down to the forces of nature. Big white boss may think he can outsmart nature but the small farmer know. Earth is our witness." This relationship to the earth meant that southern black folks, whether they were impoverished or not, knew firsthand that white supremacy, with its systemic dehumanization of blackness, was not a form of absolute power.

In that world country black folks understood that though powerful white folks could dominate and control people of color they could not control nature or divine spirit. The fundamental understanding that white folks were not gods (for if they were they could shape nature) helped imbue black folks with an oppositional sensibility. When black people migrated to urban cities, this humanizing connection with nature was severed; racism and white supremacy came to be seen as all powerful, the ultimate factors informing our fate. When this thinking was coupled with a breakdown in religiosity, a refusal to recognize the sacred in everyday life, it served the interests of white supremacist capitalist patriarchy.

Living in the agrarian South, working on the land, growing food, taught survival skills similar to those hippies sought to gain in their back-to-the-earth movements in the late sixties and early seventies. Growing up in a world where my grandparents did not hold regular jobs but made their living digging and selling fishing worms, growing food, raising chickens, I was ever mindful of an alternative to the capitalist system that destroyed nature's abundance. In that world I learned experientially the concept of interbeing, which

Buddhist monk Thich Nhat Hanh talks about as that recognition of the connectedness of all human life.

That sense of interbeing was once intimately understood by black folks in the agrarian South. Nowadays it is only those who maintain our bonds to the land, to nature, who keep our vows of living in harmony with the environment, who draw spiritual strength from nature. Reveling in nature's bounty and beauty has been one of the ways enlightened poor people in small towns all around our nations stay in touch with their essential goodness even as forces of evil, in the form of corrupt capitalism and hedonistic consumerism, work daily to strip them of their ties with nature.

Journalists from the *New York Times* who interviewed Kentucky po' rural folk getting by with scarce resources were surprised to find these citizens expressing connection to nature. In a recent article in the *Times* titled "Forget Washington: The Poor Cope Alone," reporter Evelyn Nieves shared: "People time and again said they were blessed to live in a place as beautiful as Kentucky, where the mountains are green and lush and the trees look as old as time." Maintaining intimacy gives us a concrete place of hope. It is nature that reminds time and time again that "this too will pass." To look upon a tree, or a hilly waterfall, that has stood the test of time can renew the spirit. To watch plants rise from the earth with no special tending reawakens our sense of awe and wonder.

More than ever before in our nation's history black folks must collectively renew our relationship to the earth, to our agrarian roots. For when we are forgetful and participate in the destruction and exploitation of dark earth, we collude with the domination of the earth's dark people, both here and globally. Reclaiming our history, our relationship to nature, to farming in America, and proclaiming the humanizing restorative of living in harmony with nature so that earth can be our witness is meaningful resistance.

When I leave my small flat in an urban world where nature has been so relentlessly assaulted that it is easy to forget to look at a tree, a sky, a flower emerging in a sea of trash, and go to the country, I seek renewal. To live in communion with the earth fully acknowledging nature's power with humility and grace is a practice of spiritual mindfulness that heals and restores. Making peace with the earth we make the world a place where we can be one with nature. We create and sustain environments where we can come back to ourselves, where we can return home, stand on solid ground, and be a true witness.

An Aesthetic of Blackness: Strange and Oppositional

This is the story of a house. It has been lived in by many people.
Our grandmother, Baba, made this house a living space. She was
certain that the way we lived was shaped by objects, the way we
looked at them, the way they were placed around us. She was certain
that we were shaped by space. From her I learn about aesthetics, the
yearning for beauty that she tells me is the predicament of heart that
makes our passion real. A quiltmaker, she teaches me about color.
Her house is a place where I am learning to look at things, where I
am learning how to belong in space. In rooms full of objects, crowded
with things, I am learning to recognize myself. She hands me a mir-
ror, showing me how to look. The color of wine she has made in my
cup, the beauty of the everyday. Surrounded by fields of tobacco, the
leaves braided like hair, dried and hung, circles and circles of smoke
fill the air. We string red peppers fiery hot, with thread that will not
be seen. They will hang in front of a lace curtain to catch the sun.

Look, she tells me, what the light does to color! Do you believe that space can give life, or take it away, that space has power? These are the questions she asks which frighten me. Baba dies an old woman, out of place. Her funeral is also a place to see things, to recognize myself. How can I be sad in the face of death, surrounded by so much beauty? Death, hidden in a field of tulips, wearing my face and calling my name. Baba can make them grow. Red, yellow, they surround her body like lovers in a swoon, tulips everywhere. Here a soul on fire with beauty burns and passes, a soul touched by flame. We see her leave. She has taught me how to look at the world and see beauty. She has taught me "we must learn to see."

Years ago, at an art gallery in San Francisco near the Tassajara restaurant, I saw rooms arranged by Buddhist monk Chögyam Trungpa. At a moment in my life when I had forgotten how to see, he reminds me to look. He arranges spaces. Moved by an aesthetic shaped by old beliefs. Objects are not without spirit. As living things they touch us in unimagined ways. On this path one learns that an entire room is a space to be created, a space that can reflect beauty, peace, and a harmony of being, a spiritual aesthetic. Each space is a sanctuary. I remember. Baba has taught me "we must learn to see."

Aesthetics then is more than a philosophy or theory of art and beauty; it is a way of inhabiting space, a particular location, a way of looking and becoming. It is not organic. I grew up in an ugly house. No one there considered the function of beauty or pondered the use of space. Surrounded by dead things, whose spirits had long ago vanished since they were no longer needed, that house contained a great engulfing emptiness. In that house things were not to be looked at, they were to be possessed — space was not to be created but owned — a violent anti-aesthetic. I grew up thinking about art and beauty as it existed in our lives, the lives of poor black people. Without knowing the appropriate language, I understood that advanced capitalism was affecting our capacity to see, that consumerism began to take the place of that predicament of heart that called us to yearn for beauty. Now many of us are only yearning for things. In one house I learned the place of aesthetics in the lives of agrarian poor black folks. There the lesson was that one had to understand beauty as a force to be made and imagined. Old folks shared their sense that we had come out of slavery into this free space and we had to create a world that would renew the spirit, that would make it life-giving. In that house there was a sense of history. In the other house, the one I lived in, aesthetics had no place. There the lessons were never about art or beauty, but always only to possess things. My thinking about aesthetics has been informed by the recognition of these houses: one which cultivated and celebrated an aesthetic of existence, rooted in the idea that no degree of material lack could keep one from learning how to look at the world with a critical eye, how to recognize beauty, or how to use it as a force to enhance inner well-being; the other which denied the power of abstract aestheticism. Living in that other house where we were so acutely aware of lack, so conscious of materiality, I could see in our daily life the way consumer capitalism ravaged the black poor, nurtured in us a longing for things that often subsumed our ability to recognize aesthetic worth or value.

Despite these conditions, there was in the traditional southern racially segregated black community a concern with racial uplift that continually promoted recognition of the need for artistic expressiveness and cultural production. Art was seen as intrinsically serving a political function. Whatever African-Americans created in music, dance, poetry, painting, etc., it was regarded as testimony, bearing witness, challenging racist thinking which suggested that black folks were not fully human, were uncivilized, and that the measure of this was our collective failure to create "great" art. White supremacist ideology insisted that black people, being more animal than human, lacked the capacity to feel and therefore could not engage the finer sensibilities that were the breeding ground for art. Responding to this propaganda, nineteenth-century black folks emphasized the importance of

art and cultural production, seeing it as the most effective challenge
to such assertions. Since many displaced African slaves brought to this
country an aesthetic based on the belief that beauty, especially that
created in a collective context, should be an integrated aspect of everyday life, enhancing the survival and development of community,
these ideas formed the basis of African-American aesthetics. Cultural
production and artistic expressiveness were also ways for displaced
African people to maintain connections with the past. Artistic African cultural retentions survived long after other expressions had been
lost or forgotten. Though not remembered or cherished for political
reasons, they would ultimately be evoked to counter assertions by
white supremacists and colonized black minds that there remained no
vital living bond between the culture of African-Americans and the
cultures of Africa. This historical aesthetic legacy has proved so powerful that consumer capitalism has not been able to completely destroy
artistic production in underclass black communities.

Even though the house where I lived was ugly, it was a place
where I could and did create art. I painted, I wrote poetry. Though
it was an environment more concerned with practical reality than
art, these aspirations were encouraged. In an interview in *Callaloo*,
painter Lois Mailou Jones describes the tremendous support she
received from black folks: "Well, I began with art at a very early
stage in my life. As a child, I was always drawing. I loved color. My
mother and father, realizing that I had talent, gave me an excellent
supply of crayons and pencils and paper — and encouraged me."
Poor black parents saw artistic cultural production as crucial to the
struggle against racism, but they were also cognizant of the link between creating art and pleasure. Art was necessary to bring delight,
pleasure, and beauty into lives that were hard, that were materially
deprived. It mediated the harsh conditions of poverty and servitude.
Art was also a way to escape one's plight. Protestant black churches
emphasized the parable of the talents, and commitment to spirituality also meant appreciating one's talents and using them. In our

church if someone could sing or play the piano and they did not offer these talents to the community, they were admonished.

Performance arts — dance, music, and theater — were the most accessible ways to express creativity. Making and listening to black music, both secular and sacred, was one of the ways black folks developed an aesthetic. It was not an aesthetic documented in writing, but it did inform cultural production. Analyzing the role of the "talent show" in segregated black communities, which was truly the community-based way to support and promote cultural production, would reveal much about the place of aesthetics in traditional black life. It was both a place for collective display of artistry and a place for the development of aesthetic criteria. I cite this information to place African-American concern with aesthetics in a historical framework that shows a continuity of concern. It is often assumed that black folks first began to articulate an interest in aesthetics during the sixties. Privileged black folks in the nineteenth and early twentieth centuries were often, like their white counterparts, obsessed with notions of "high art." Significantly, one of the important dimensions of the artistic movement among black people, most often talked about as the Harlem Renaissance, was the call for an appreciation of popular forms. Like other periods of intense focus on the arts in African-American culture, it called attention to forms of artistic expression that were simply passing away because they were not valued in the context of a conventional aesthetic focusing on "high art." Often African-American intellectual elites appropriated these forms, reshaping them in ways suited to different locations. Certainly the spiritual as it was sung by Paul Robeson at concerts in Europe was an aspect of African-American folk culture evoked in a context far removed from small, hot, southern church services, where poor black folks gathered in religious ecstasy. Celebration of popular forms ensured their survival, kept them as a legacy to be passed on, even as they were altered and transformed by the interplay of varied cultural forces.

Conscious articulation of a "black aesthetic" as it was constructed by African-American artists and critics in the sixties and early seventies was an effort to forge an unbreakable link between artistic production and revolutionary politics. Writing about the interconnectedness of art and politics in the essay "Frida Kahlo and Tina Modottit," Laura Mulvey describes the way an artistic avant-grade

> was able to use popular form not as a means of communication but as a means of constructing a mythic past whose effectiveness could be felt in the present. Thereby it brought itself into line with revolutionary impetus towards constructing the mythic past of the nation.

A similar trend emerged in African-American art as painters, writers, musicians worked to imaginatively evoke black nationhood, a homeland, re-creating bonds with an African past while simultaneously evoking a mythic nation to be born in exile. During this time Larry Neal declared the Black Arts Movement to be "the cultural arm of the black revolution." Art was to serve black people in the struggle for liberation. It was to call for and inspire resistance. One of the major voices of the black aesthetic movement, Maulana Karenga, in his *Thesis on Black Cultural Nationalism*, taught that art should be functional, collective, and committed.

The black aesthetic movement was fundamentally essentialist. Characterized by an inversion of the "us" and "them" dichotomy, it inverted conventional ways of thinking about otherness in ways that suggested that everything black was good and everything white bad. In his introduction to the anthology *Black Fire*, Larry Neal set the terms of the movement, dismissing work by black artists which did not emerge from black power movement:

> A revolutionary art is being expressed today. The anguish and aimlessness that attended our great artists of the forties

and fifties and which drove most of them to early graves, to dissipation and dissolution, is over. Misguided by white cultural references (the models the culture sets for its individuals), and the incongruity of these models with black reality, men like Bird were driven to willful self-destruction. There was no program. And the reality-model was incongruous. It was a white reality-model. If Bird had had a black reality-model, it might have been different ... In Bird's case, there was a dichotomy between his genius and the society. But that he couldn't find the adequate model of being was the tragic part of the whole thing.

Links between black cultural nationalism and revolutionary politics led ultimately to the subordination of art to politics. Rather than serving as a catalyst promoting diverse artistic expression, the Black Arts Movement began to dismiss all forms of cultural production by African-Americans that did not conform to movement criteria. Often this led to aesthetic judgments that did not allow for recognition of multiple black experiences or the complexity of black life, as in the case of Neal's critical interpretation of jazz musician Charlie Parker's fate. Clearly, the problems facing Parker were not simply aesthetic concerns, and they could not have been resolved by art of critical theories about the nature of black artistic production. Ironically, in many of its aesthetic practices the Black Arts Movement was based on the notion that a people's art, cultural production for the masses, could not be either complex, abstract, or diverse in style, form, content, etc.

Despite its limitations, the Black Arts Movement provided useful critique based on radical questioning of the place and meaning of aesthetics for black artistic production. The movement's insistence that all art is political, that an ethical dimension should inform cultural production, as well as the encouragement of an aesthetic which did not separate habits of being from artistic production, were important to black thinkers concerned with strategies of decolonization. Unfortunately, these positive aspects of the black aesthetic movement should have led to the formation of critical space where there could have been more open discussion of the relevance of cultural production to black liberation struggle. Ironically, even though the Black Arts Movement insisted that it represented a break from white western traditions, much of its philosophical underpinning re-inscribed prevailing notions about the relationship between art and mass culture. The assumption that naturalism or realism was more accessible to a mass audience than abstraction was certainly not a revolutionary position. Indeed the paradigms for artistic creation offered by the Black Arts Movement were most often restrictive and disempowering. They stripped many artists of creative agency by dismissing and devaluing their work because it was either too abstract or did not overtly address a radical politic. Writing about socialist attitudes towards art and politics in *Art and Revolution*, John Berger suggests that the relationship between art and political propaganda is often confused in the radical or revolutionary context. This was often the case in the Black Arts Movement. While Berger willingly accepts the truism "that all works of art exercise an ideological influence — even works by artists who profess to have no interest outside art," he critiques the idea that simplicity of form or content necessarily promotes critical political consciousness or leads to the development of a meaningful revolutionary art. His words of caution should be heeded by those who would revive a prescriptive black aesthetic that limits freedom and restricts artistic development. Speaking against a prescriptive aesthetic, Berger writes:

> When the experience is "offered up," it is not expected to be in any way transformed. Its apotheosis should be instant, and as it were invisible. The artistic process is taken for granted: it always remains exterior to the spectator's experience. It is no more than the supplied vehicle in which experience is

placed so that it may arrive safely at a kind of cultural terminus. Just as academicism reduces the process of art to an apparatus for artists, it reduces it to a vehicle for the spectator. There is absolutely no dialectic between experience and expression, between experience and its formulations.

The black aesthetic movement was a self-conscious articulation by many of a deep fear that the power of art resides in its potential to transgress boundaries.

Many African-American artists retreated from black cultural nationalism into a retrogressive posture where they suggested there were no links between art and politics, evoking outmoded notions of art as transcendent and pure to defend their position. This was another step backwards. There was no meaningful attempt to counter the black aesthetic with conceptual criteria for creating and evaluating art which would simultaneously acknowledge its ideological content even as it allowed for expansive notions of artistic freedom. Overall the impact of these two movements, black aesthetics and its opponents, was a stifling of artistic production by African-Americans in practically every medium with the exception of music. Significantly, avant-garde jazz musicians, grappling with artistic expressivity that demanded experimentation, resisted restrictive mandates about their work, whether they were imposed by a white public saying their work was not really music or a black public which wanted to see more overt links between that work and political struggle.

To re-open the creative space that much of the black aesthetic movement closed down, it seems vital for those involved in contemporary black arts to engage in a revitalized discussion of aesthetics. Critical theories about cultural production, about aesthetics, continue to confine and restrict black artists, and passive withdrawal from a discussion of aesthetics is a useless response. To suggest, as Clyde Taylor does in his essay "We Don't Need Another Hero: Anti-Theses On Aesthetics," that the failure of black aesthetics or the development of white western theorizing on the subject should negate all African-American concern with the issue is to once again repeat an essentialist project that does not enable or promote artistic growth. An African-American discourse on aesthetics need not begin with white western traditions and it need not be prescriptive. Cultural decolonization does not happen solely by repudiating all that appears to maintain connection with the colonizing culture. It is really important to dispel the notion that white western culture is "the" location where a discussion of aesthetics emerged, as Taylor suggests; it is only one location.

Progressive African-Americans concerned with the future of our cultural production seek to critically conceptualize a radical aesthetic that does not negate the powerful place of theory as both that force which sets up criteria for aesthetic judgment and as vital grounding that helps make certain work possible, particularly expressive work that is transgressive and oppositional. Hal Foster's comments on the importance of an anti-aesthetic in the essay "Postmodernism: A Preface" present a useful paradigm African-Americans can employ to interrogate modernist notions of aesthetics without negating the discourse on aesthetics. Foster proposes this paradigm to critically question "the idea that aesthetic experience exists apart, without 'purpose,' all but beyond history, or that art can now affect a world at once (inter) subjective, concrete, and universal — a symbolic totality." Taking the position that an anti-aesthetic "signals a practice, cross-disciplinary in nature, that is sensitive to cultural forms engaged in a politic (e.g., feminist art) or rooted in a vernacular — that is, to forms that deny the idea of a privileged aesthetic realm," Foster opens up the possibility that work by marginalized groups can have a greater audience and impact. Working from a base where difference and otherness are acknowledged as forces that intervene in western theorizing about aesthetics to reformulate and transform the discussion, African-Americans are empowered to break with old ways of seeing reality that suggest there is only one audience for our work and only one aesthetic measure of its value. Moving away from narrow cultural nationalism, one leaves behind as well racist assumptions that cultural productions by black people can only have "authentic" significance and meaning for a black audience. Black artists concerned with producing work that embodies and reflects a liberatory politic know that an important part of any decolonization process is critical intervention and interrogation of existing repressive and dominating structures. African-American critics and/or artists who speak about our need to engage in ongoing dialogue with dominant discourses always risk being dismissed as assimilationist. There is a grave difference between that engagement with white culture which seeks to deconstruct, demystify, challenge, and transform and gestures of collaboration and complicity. We cannot participate in dialogue that is the mark of freedom and critical agency if we dismiss all work emerging from white western traditions. The assumption that the crisis of African-Americans should or can only be addressed by us must also be interrogated. Much of what threatens our collective wellbeing is the product of dominating structures. Racism is a white issue as much as it is a black one.

Contemporary intellectual engagement with issues of "otherness and difference" manifest in literary critique, cultural studies, feminist theory, and black studies indicates that there is a growing body of work that can provide and promote critical dialogue and debate across boundaries of class, race, and gender. These circumstances, coupled with a focus on pluralism at the level of social and public policy, are creating a cultural climate where it is possible to interrogate the idea that difference is synonymous with lack and deprivation, and simultaneously call for critical re-thinking of aesthetics. Retrospective examination of the repressive impact a prescriptive black aesthetic had on black cultural production should serve as a cautionary model for African-Americans. There can never be one critical paradigm for the evaluation of artistic work. In part, a radical aesthetic acknowledges that we are constantly changing positions, locations, that our needs and concerns vary, that these diverse directions must correspond with shifts in critical thinking. Narrow limiting aesthetics within black communities tend to place innovative black artistry on the margins. Often this work receives little or no attention. Whenever black artists work in ways that are transgressive, we are seen as suspect, by our group and by the dominant culture. Rethinking aesthetic principles could lead to the development of a critical standpoint that promotes and encourages various modes of artistic and cultural production.

As artist and critic, I find compelling a radical aesthetic that seeks to uncover and restore links between art and revolutionary politics, particularly black liberation struggle, while offering an expansive critical foundation for aesthetic evaluation. Concern for the contemporary plight of black people necessitates that I interrogate my work to see if it functions as a force that promotes the development of critical consciousness and resistance movement. I remain passionately committed to an aesthetic that focuses on the purpose and function of beauty, of artistry in everyday life, especially the lives of poor people, one that seeks to explore and celebrate the connection between our capacity to engage in critical resistance and our ability to experience pleasure and beauty. I want to create work that shares with an audience, particularly oppressed and marginalized groups, the sense of agency artistry offers, the empowerment. I want to share the aesthetic inheritance handed down to me by my grandmother and generations of black ancestors, whose ways of thinking about the issue have been globally shaped in the African diaspora and informed by the experience of exile and domination. I want to reiterate the message that "we must learn to see." Seeing here is meant metaphysically as heightened awareness and understanding, the intensification of one's capacity to experience reality through the realm of the senses.

Remembering the houses of my childhood, I see how deeply my concern with aesthetics was shaped by black women who were

fashioning an aesthetic of being, struggling to create an oppositional world view for their children, working with space to make
it livable. Baba, my grandmother, could not read or write. She did
not inherit her contemplative preoccupation with aesthetics from
a white western literary tradition. She was poor all her life. Her
memory stands as a challenge to intellectuals, especially those on
the left, who assume that the capacity to think critically, in abstract
concepts, to be theoretical, is a function of class and educational
privilege. Contemporary intellectuals committed to progressive
politics must be reminded again and again that the capacity to name
something (particularly in writing terms like aesthetics, postmodernism, deconstruction, etc.) is not synonymous with the creation
or ownership of the condition or circumstance to which such terms
may refer.

Many underclass black people who do not know conventional
academic theoretical language are thinking critically about aesthetics. The richness of their thoughts is rarely documented in books.
Innovative African-American artists have rarely documented their
process, their critical thinking on the subject of aesthetics. Accounts
of the theories that inform their work are necessary and essential;
hence my concern with opposing any standpoint that devalues this
critical project. Certainly many of the revolutionary, visionary critical perspectives on music that were inherent to John Coltrane's oppositional aesthetics and his cultural production will never be shared
because they were not fully documented. Such tragic loss retards the
development of reflective work by African-Americans on aesthetics
that is linked to enabling politics. We must not deny the way aesthetics serves as the foundation for emerging visions. It is, for some
of us, critical space that inspires and encourages artistic endeavor.
The ways we interpret that space and inhabit it differ.

As a grown black woman, a guest in my mother's house, I explain
that my interior landscape is informed by minimalism, that I cannot
live in a space filled with too many things. My grandmother's house

is only inhabited by ghosts and can no longer shelter or rescue me. Boldly I declare that I am a minimalist. My sisters repeat this word with the kind of glee that makes us laugh, as we celebrate together that particular way language, and the "meaning" of words is transformed when they fall from the hierarchical space they inhabit in certain locations (the predominantly white university setting) into the mouths of vernacular culture and speech, into underclass blackness, segregated communities where there is much illiteracy. Who can say what will happen to this word "minimalist"? Who knows how it will be changed, re-fashioned by the thick patois that is our southern black tongue? This experience cannot be written. Even if I attempt description, it will never convey process.

One of my five sisters wants to know how it is I come to think about these things, about houses, and space. She does not remember long conversations with Baba. She remembers her house as an ugly place, crowded with objects. My memories fascinate her. She listens with astonishment as I describe the shadows in Baba's house and what they meant to me, the way the moon entered an upstairs window and created new ways for me to see dark and light. After reading Tanizaki's essay on aesthetics "In Praise of Shadows," I tell this sister in a late-night conversation that I am learning to think about blackness in a new way. Tanizaki speaks of seeing beauty in darkness and shares this moment of insight: "The quality that we call beauty, however, must always grow from the realities of life, and our ancestors, forced to live in dark rooms, presently came to discover beauty in shadows, ultimately to guide shadows towards beauty's end." My sister has skin darker than mine. We think about our skin as a dark room, a place of shadows. We talk often about color politics and the ways racism has created an aesthetic that wounds us, a way of thinking about beauty that hurts. In the shadows of late night, we talk about the need to see darkness differently, to talk about it in a new way. In that space of shadows we long for an aesthetic of blackness — strange and oppositional.

Inspired Eccentricity

There are family members you try to forget and ones that you always remember, that you can't stop talking about. They may be dead — long gone — but their presence lingers and you have to share who they were and who they still are with the world. You want everyone to know them as you did, to love them as you did.

All my life I have remained enchanted by the presence of my mother's parents, Sarah and Gus Oldham. When I was a child they were already old. I did not see that then, though. They were Baba and Daddy Gus, together for more than seventy years at the time of his death. Their marriage fascinated me. They were strangers and lovers — two eccentrics who created their own world.

More than any other family members, together they gave me a worldview that sustained me during a difficult and painful childhood. Reflecting on the eclectic writer I have become, I see in myself a mixture of these two very different but equally powerful figures from my childhood. Baba was tall, her skin so white and her hair so jet black and straight that she could have easily "passed" denying all traces of blackness. Yet the man she married was short and dark and sometimes his skin looked like the color of soot from burning coal. In our childhood the fireplaces burned coal. It was bright heat luminous and fierce. If you got too close it could burn you.

Together Baba and Daddy Gus generated a hot heat. He was a man of few words, deeply committed to silence — so much so that it was like a religion to him. When he spoke you could hardly hear what he said. Baba was just the opposite. Smoking an abundance of cigarettes a day, she talked endlessly. She preached. She yelled. She fussed. Often her vitriolic rage would heap itself on Daddy Gus who would sit calmly in his chair by the stove as calm and still as the Buddha sits. And when he had enough of her words, he would reach for his hat and walk.

Neither Baba nor Daddy Gus drove cars. Rarely did they ride in them. They preferred walking. And even then their styles were different. He moved slow, as though carrying a great weight: she with her tall, lean, boyish frame moved swiftly, as though there was never time to waste. Their one agreed-upon passion was fishing. Though they did not do even that together. They lived close but they created separate worlds.

In a big two-story wood frame house with lots of rooms they constructed a world that could contain their separate and distinct personalities. As children one of the first things we noticed about our grandparents was that they did not sleep in the same room. This arrangement was contrary to everything we understood about marriage. While Mama never wanted to talk about their separate worlds, Baba would tell you in a minute that Daddy Gus was nasty, that he smelled like tobacco juice, that he did not wash enough, that there was no way she would want him in her bed. And while he would say nothing nasty about her, he would merely say why would he want to share somebody else's bed when he could have his own bed to himself, with no one to complain about anything.

I loved my granddaddy's smells. Always, they filled my nostrils with the scent of happiness. It was sheer ecstasy for me to be allowed into his inner sanctum. His room was a small Van Gogh — like space off from the living room. There was no door. Old-fashioned curtains were the only attempt at privacy. Usually the curtains were closed. His room reeked of tobacco. There were treasures everywhere in that small room. As a younger man Daddy Gus did odd jobs, and sometimes even in his old age he would do a chore for some needy lady. As he went about his work, he would pick up found objects, scraps. All these objects would lie about his room, on the dresser, on the table near his bed. Unlike all other grown-ups he never cared about children looking through his things. Anything we wanted he gave to us.

Daddy Gus collected beautiful wooden cigar boxes. They held lots of the important stuff — the treasures. He had tons of little diaries that he made notes in. He gave me my first wallet: my first teeny little book to write in, my first beautiful pen, which did not write for long, but it was still a found and shared treasure. When I would lie on his bed or sit close to him, sometimes just standing near, I would feel all the pain and anxiety of my troubled childhood leave me. His spirit was calm. He gave me the unconditional love I longed for.

"Too calm," his grown-up children thought. That's why he had let this old woman rule him, my cousin BoBo would say. Even as children we knew that grown-ups felt sorry for Daddy Gus. At times his sons seemed to look upon him as not a "real man." His refusal to fight in wars was another sign to them of weakness. It was my grandfather who taught me to oppose war. They saw him as a man controlled by the whims of others, by this tall, strident, demanding woman he had married. I saw him as a man of profound beliefs, a man of integrity. When he heard their put-downs — for they talked on and on about his laziness — he merely muttered that he had no use for them. He was not gonna let anybody tell him what to do with his life.

Daddy Gus was a devout believer, a deacon at his church; he was one of the right-hand men of God. At church, everyone admired his calmness. Baba had no use for church. She liked nothing better than to tell us all the ways it was one big hypocritical place: "Why, I can find God anywhere I want to — I do not need a church." Indeed, when my grandmother died, her funeral could not take place in a church, for she had never belonged. Her refusal to attend church bothered some of her daughters, for they thought she was sinning against God, setting a bad example for the children. We were not supposed to listen when she began to damn the church and everybody in it.

Baba loved to "cuss." There was no bad word she was not willing to say. The improvisational manner in which she would string those words together was awesome. It was the goddamn sons of bitches who thought that they could fuck with her when they could just kiss her black ass. A woman of strong words and powerful metaphors, she could not read or write. She lived in the power of language. Her favorite sayings were a prelude for storytelling. It was she who told me, "Play with a puppy, he'll lick you in the mouth." When I heard this saying, I knew what was coming — a long polemic about not letting folks get too close, 'cause they will mess with you.

Baba loved to tell her stories. And I loved to hear them. She called me Glory. And in the midst of her storytelling she would pause to say, "Glory, are ya listenin'? Do you understand what I'm telling ya?" Sometimes I would have to repeat the lessons I had learned. Sometimes I was not able to get it right and she would start again. When Mama felt I was learning too much craziness "over home" (that is what we called Baba's house), my visits were curtailed. As I moved into my teens I learned to keep to myself all the wisdom of the old ways I picked up over home.

Baba was an incredible quiltmaker, but by the time I was old enough to really understand her work, to see its beauty, she was already having difficulty with her eyesight. She could not sew as much as in the old days, when her work was on everybody's bed. Unwilling to throw anything away, she loved to make crazy quilts, 'cause they allowed every scrap to be used. Although she would one day order patterns and make perfect quilts with colors that went together, she always collected scraps.

Long before I read Virginia Woolf's *A Room of One's Own* I learned from Baba that a woman needed her own space to work. She had a huge room for her quilting. Like every other space in the private world she created upstairs, it had her treasures, an endless array of hatboxes, feathers, and trunks filled with old clothes she had held on to. In room after room there were feather tick mattresses; when they were pulled back, the wooden slats of the bed were revealed, lined with exquisite hand-sewn quilts.

In all these trunks, in crevices and drawers were braided tobacco leaves to keep away moths and other insects. A really hot summer could make cloth sweat, and stains from tobacco juice would end up on quilts no one had ever used. When I was a young child, a quilt my grandmother had made kept me warm, was my solace and comfort. Even though Mama protested when I dragged that old raggedy quilt from Kentucky to Stanford, I knew I needed that bit of the South, of Baba's world, to sustain me.

Like Daddy Gus, she was a woman of her word. She liked to declare with pride. "I mean what I say and I say what I mean." "Glory," she would tell me, "nobody is better than their word — if you can't keep ya word you ain't worth nothin' in this world." She would stop speaking to folk over the breaking of their word, over lies. Our Mama was not given to loud speech or confrontation. I learned all those things from Baba — "to stand up and speak up" and not to "give a good goddamn" what folk who "ain't got a pot to pee in" think. My parents were concerned with their image in the world. It was pure blasphemy for Baba to teach that it did not matter what other folks thought — "Ya have to be right with yaself in ya own heart — that's all that matters." Baba taught me to listen to my heart — to follow it. From her we learned as small children to remember our dreams in the night and to share them when we

awakened. They would be interpreted by her. She taught us to listen
to the knowledge in dreams. Mama would say this was all nonsense,
but she too was known to ask the meaning of a dream.

In their own way my grandparents were rebels, deeply committed to radical individualism. I learned how to be myself from
them. Mama hated this. She thought it was important to be liked,
to conform. She had hated growing up in such an eccentric, other-
wordly household. This world where folks made their own wine,
their own butter, their own soap; where chickens were raised, and
huge gardens were grown for canning everything. This was the
world Mama wanted to leave behind. She wanted store-bought
things.

Baba lived in another time, a time when all things were produced
in the individual household. Everything the family needed was made
at home. She loved to tell me stories about learning to trap animals, to
skin, to soak possum and coon in brine, to fry up a fresh rabbit. Though
a total woman of the outdoors who could shoot and trap as good as any
man, she still believed every woman should sew — she made her first
quilt as a girl. In her world, women were as strong as men because they
had to be. She had grown up in the country and knew that country
ways were the best ways to live. Boasting about being able to do anything that a man could do and better, this woman who could not read
or write was confident about her place in the universe.

My sense of aesthetics came from her. She taught me to really
look at things, to see underneath the surface, to see the different
shades of red in the peppers she had dried and hung in the kitchen
sunlight. The beauty of the ordinary, the everyday, was her feast of
light. While she had no use for the treasures in my granddaddy's
world, he too taught me to look for the living spirit in things — the
things that are cast away but still need to be touched and cared for.
Picking up a found object he would tell me its story or tell me how
he was planning to give it life again.

Connected in spirit but so far apart in the life of everydayness.
Baba and Daddy Gus were rarely civil to each other. Every shared

talk begun with goodwill ended in disagreement and contestation. Everyone knew Baba just loved to fuss. She liked a good war of words. And she was comfortable using words to sting and hurt, to punish. When words would not do the job, she could reach for the strap, a long piece of black leather that would leave tiny imprints on the flesh.

There was no violence in Daddy Gus. Mama shared that he had always been that way, a calm and gentle man, full of tenderness. I remember clinging to his tenderness when nothing I did was right in my mother's eyes, when I was constantly punished. Baba was not an ally. She advocated harsh punishment. She had no use for children who would not obey. She was never ever affectionate. When we entered her house, we gave her a kiss in greeting and that was it. With Daddy Gus we could cuddle, linger in his arms, give as many kisses as desired. His arms and heart were always open.

In the back of their house were fruit trees, chicken coops, and gardens, and in the front were flowers. Baba could make anything grow. And she knew all about herbs and roots. Her home remedies healed our childhood sicknesses. Of course she thought it crazy for anyone to go to a doctor when she could tell them just what they needed. All these things she had learned from her mother, Bell Blair Hooks, whose name I would choose as my pen name. Everyone agreed that I had the temperament of this great-grandmother I would not remember. She was a sharp-tongued woman. Or so they said. And it was believed I had inherited my way with words from her.

Families do that. They chart psychic genealogies that often overlook what is right before our eyes. I may have inherited my great-grandmother Bell Hooks' way with words, but I learned to use those words listening to my grandmother. I learned to be courageous by seeing her act without fear. I learned to risk because she was daring. Home and family were her world. While my grandfather journeyed downtown, visited at other folks houses, went to church, and conducted affairs in the world, Baba rarely left home.

There was nothing in the world she needed. Things out there violated her spirit.

As a child I had no sense of what it would mean to live a life spanning so many generations unable to read or write. To me Baba was a woman of power. That she would have been extraordinarily powerless in a world beyond 1200 Broad Street was a thought that never entered my mind. I believed that she stayed home because it was the place she liked best. Just as Daddy Gus seemed to need to walk — to roam.

After his death it was easier to see the ways that they complemented and completed each other. For suddenly, without him as a silent backdrop, Baba's spirit was diminished. Something in her was forever lonely and could not find solace. When she died, tulips, her favorite flower, surrounded her. The preacher told us that her death was not an occasion for grief, for "it is hard to live in a world where your choicest friends are gone." Daddy Gus was the companion she missed most. His presence had always been the mirror of memory. Without it there was so much that could not be shared. There was no witness.

Seeing their life together, I learned that it was possible for women and men to fashion households arranged around their own needs. Power was shared. When there was an imbalance, Baba ruled the day. It seemed utterly alien to me to learn about black women and men not making families and homes together. I had not been raised in a world of absent men. One day I knew I would fashion a life using the patterns I inherited from Baba and Daddy Gus. I keep treasures in my cigar box, which still smells after all these years. The quilt that covered me as a child remains, full of ink stains and faded colors. In my trunks are braided tobacco leaves, taken from over home. They keep evil away — keep bad spirits from crossing the threshold. Like the ancestors they guard and protect.

13

A Place Where the
Soul Can Rest

Street corners have always been space that has belonged to men — patriarchal territory. The feminist movement did not change that. Just as it was not powerful enough to take back the night and make the dark a safe place for women to lurk, roam, and meander at will, it was not able to change the ethos of the street corner — gender equality in the workplace, yes, but the street corner turns every woman who dares lurk into a body selling herself, a body looking for drugs, a body going down. A female lurking, lingering, lounging on a street corner is seen by everyone, looked at, observed. Whether she wants to be or not she is prey for the predator, for the Man, be he pimp, police, or just passerby. In cities women have no outdoor territory to occupy. They must be endlessly moving or enclosed. They must have a destination. They cannot loiter or linger.

Verandas and porches were made for females to have outdoor space to occupy. They are a common feature of southern living. Before air-conditioning cooled every hot space the porch was the summertime place, the place everyone flocked to in the early mornings and in the late nights. In our Kentucky world of poor southern black neighborhoods of shotgun houses and clapboard houses, a porch was a sign of living a life without shame. To come out on the porch was to see and be seen, to have nothing to hide. It signaled a willingness to be known. Oftentimes the shacks of the destitute were places where inhabitants walked outside straight into dust and dirt — there was neither time nor money to make a porch.

The porches of my upbringing were places of fellowship — outside space women occupied while men were away, working or on street corners. To sit on one's porch meant chores were done — the house was cleaned, food prepared. Or if you were rich enough and the proud possessor of a veranda, it was the place of your repose while the house-keeper or maid finished your cleaning. As children we needed permission to sit on the porch, to reside if only for a time, in that place of leisure and rest. The first house we lived in had no porch. A cinder-block dwelling made for working men to live in while they searched the earth for oil outside city limits, it was designed to be a waiting place, a place for folks determined to move up and on — a place in the wilderness. In the wilderness there were no neighbors to wave at or chat with or simply to holler at and know their presence by the slamming of doors as one journeyed in and out. A home without neighbors surely did not require a porch, just narrow steps to carry inhabitants in and out.

When we moved away from the wilderness, when we moved up, our journey of improved circumstances took us to a wood frame house with upstairs and downstairs. Our new beginning was grand: we moved to a place with not one but three porches — a front porch, a side porch, and a back porch. The side porch was a place where folks could sleep when the heat of the day had cooled off. Taking one's dreams outside made the dark feel safe. And in that safeness, a woman, a child — girl or boy — could linger. Side porches were places for secret meetings, places where intimate callers could come and go without being seen, spend time without anyone knowing how long they stayed. After a year of living with a side porch and six teenaged girls, Daddy sheetrocked, made walls, blocked up the door so that it became our brother's room, an enclosed space with no window to the outside.

We sat on the back porch and did chores like picking walnuts, shucking corn, and cleaning fish, when Baba, Mama's mama, and the rest had a good fishing day, when black farmers brought the fruit of their labor into the city. Our back porch was tiny. It could not hold all of us. And so it was a limited place of fellowship. As a child I felt more comfortable there, unobserved, able to have my child's musings, my day-dreams, without the interruptions of folks passing by and saying a word or two, without folks coming up to sit a spell. At Mr. Porter's house (he was the old man who lived and died there before we moved in) there was feeling of eternity, of timelessness. He had imprinted on the soul of this house his flavor, the taste and scent of a long-lived life. We honored that by using his name when talking about the house on First Street.

To our patriarchal dad, Mr. V, the porch was a danger zone — as in his sexist mindset all feminine space was designated dangerous, a threat. A strange man walking on Mr. V's porch was setting himself up to be a possible target: walking onto the porch, into an inner feminine sanctum, was in the eyes of any patriarch just the same as raping another man's woman. And we were all of us — mother, daughters — owned by our father. Like any patriarch would, he reminded us from time to time whose house we lived in — a house where women had no rights but could indeed claim the porch — colonize it and turn it into a place where men could look but not touch — a place that did not interest our father, a place where he did not sit. Indeed, our daddy always acted as though he hated the porch. Often when he came home from work he entered through the back door, making his territory, taking us unaware.

We learned that it was best not to be seen on the porch often when he walked up the sidewalk after a long day's work. We knew

our place: it was inside, making the world comfortable for the patriarch, preparing ourselves to bow and serve — not literally to bow,
but to subordinate our beings. And we did. No wonder then that we
loved the porch, longed to move outside the protected patriarchal
space of that house that was in its own way a prison.

Like so much else ruined by patriarchal rage, so much other female space damaged, our father the patriarch took the porch from
us one intensely hot summer night. Returning home from work in
a jealous rage, he started ranting the moment he hit the sidewalk
leading up to the steps, using threatening, ugly words. We were
all females there on that porch, parting our bodies like waves in
the sea so that Mama could be pushed by hurting hands, pushed
through the front door, pushed into the house, where his threats
to kill and kill again would not be heard by the neighbors. This
trauma of male violence took my teenage years and smothered
them in the arms of a deep and abiding grief — took away the
female fellowship, the freedom of days and nights sitting on the
porch.

Trapped in the interstices of patriarchal gender warfare, we
stayed off the porch, for fear that just any innocent male approaching
would be seen by our father and set off crazy rage. Coming in from
the outside I would see at a distance the forlorn look of a decimated
space, its life energy gone and its heart left lonely. Mama and Daddy
mended the wounded places severed by rage, maintaining their intimate bond. They moved away from Mr. Porter's house into a small
new wood frame structure, a house without a porch, and even when
a small one was added, it was not a porch for sitting, just a place for
standing. Maybe this space relieved Dad's anxiety about the dangerous feminine, about female power.

Surely our father, like all good patriarchs, sensed that the porch as
female gathering place represented in some vital way a threat to the
male dominator's hold on the household. The porch as liminal space,
standing between the house and the world of sidewalks and streets,
was symbolically a threshold. Crossing it opened up the possibility of

change. Women and children on the porch could begin to interpret the outside world on terms different from the received knowledge gleaned in the patriarchal household. The porch had no master; even our father could not conquer it. Porches could be abandoned but they could not be taken over, occupied by any one group to the exclusion of others.

A democratic meeting place, capable of containing folks from various walks of life, with diverse perspectives, the porch was free-floating space, anchored only by the porch swing, and even that was a symbol of potential pleasure. The swing hinted at the underlying desire to move freely, to be transported. A symbol of play, it captured the continued longing for childhood, holding us back in time, entrancing us, hypnotizing us with its back-and-forth motion. The porch swing was a place where intimacies could be forged, desire arising in the moment of closeness swings made possible.

In the days of my girlhood, when everyone sat on their porches, usually on their swings, it was the way we all became acquainted with one another, the way we created community. In M. Scott Peck's work on community-making and peace, *The Different Drum*, he explains that true community is always integrated and that "genuine community is always characterized by integrity." The integrity that emerged in our segregated communities as I was growing up was based on the cultivation of civility, of respect for others and acknowledgment of their presence. Walking by someone's house, seeing them on their porch, and failing to speak was to go against the tenets of the community. Now and then, I or my siblings would be bold enough to assume we could ignore the practice of civility, which included learning respect for one's elders, and strut by folks' houses and not speak. By the time we reached home, Mama would have received a call about our failure to show courtesy and respect. She would make us take our walk again and perform the necessary ritual of speaking to our neighbors who were sitting on their porches.

In *A World Waiting to Be Born: The Search for Civility*, M. Scott Peck extends his conversation on making community to include the practice of civility. Growing up in the segregated South, I was raised to believe in the importance of being civil. This was more than just a recognition of the need to be polite, of having good manners; it was a demand that I and my siblings remain constantly aware of our interconnectedness and interdependency on all the folk around us. The lessons learned by seeing one's neighbors on their porches and stopping to chat with them, or just to speak courteously, was a valuable way to honor our connectedness. Peck shares the insight that civility is consciously motivated and essentially an ethical practice. By practicing civility we remind ourselves, he writes, that "each and every human being — you, every friend, every stranger, every foreigner is precious." The etiquette of civility then is far more than the performance of manners: it includes an understanding of the deeper psychoanalytic relationship to recognition as that which makes us subjects to one another rather than objects.

African Americans have a long history of struggling to stand as subjects in a place where the dehumanizing impact of racism works continually to make us objects. In our small-town segregated world, we lived in communities of resistance, where even the small everyday gesture of porch sitting was linked to humanization. Racist white folks often felt extreme ire when observing a group of black folks gathered on a porch. They used derogatory phrases like "porch monkey" both to express contempt and to once again conjure up the racist iconography linking blackness to nature, to animals in the wild. As a revolutionary threshold between home and street, the porch as liminal space could also then be a place of anti-racist resistance. While white folk could interpret at will the actions of a black person on the street, the black person or persons gathered on a porch defied such interpretation. The racist eye could only watch, yet never truly know, what was taking place on porches among black folk.

I was a little girl in a segregated world when I first learned that there were white people who saw black people as less than animals. Sitting on the porch, my siblings and I would watch white folks bring home their servants, the maids and cooks who toiled to make ·their lives comfortable. These black servants were always relegated to the back seat. Next to the white drivers in the front would be the dog and in the back seat the black worker. Just seeing this taught me much about the interconnectedness of race and class. I often wondered how the black worker felt when it came time to come home and the dog would be placed in front, where racism and white supremacy had decreed no black person could ride. Although just a child, witnessing this act of domination, I understood that the workers must have felt shamed, because they never looked out the window; they never acknowledged a world beyond that moving car.

It was as though they were riding home in a trance — closing everything out was a way to block out the shaming feelings. Silent shadows slouched in the back seats of fancy cars, lone grown-up workers never turned their gaze toward the porch where "liberated" black folks could be seen hanging together. I was the girl they did not see, sitting in the swing, who felt their pain and wanted to make it better. And I would sit there and swing, going back and forth to the dreaming rhythm of a life where black folks would live free from fear.

Leaving racialized fear behind, I left the rhythm of porch swings, of hot nights filled with caring bodies and laughter lighting the dark like June bugs. To the West Coast I went to educate myself, away from the lazy apartheid of a Jim Crow that had been legislated away but was still nowhere near gone, to the North where I could become the intellectual the South back then had told me I could not be. But like the black folks anthropologist Carol Stack writes about, who flee the North and go South again, yearning for a life they fear is passing them by, I too returned home. To any southerner who has ever loved the South, it is always and eternally home. From birth onward we breathed in its seductive heady scent, and it is the air that truly comforts. From birth onward as southerners we were seduced and imprinted by glimpses of a civic life expressed in communion not found elsewhere. That life was embodied for me in the world of the porch.

Looking for a home in the new South, that is, the place where Jim Crow finds its accepted expression in crude acting out, I entered a real-estate culture where material profit was stronger than the urge to keep neighborhoods and races pure. Seeking to live near water, where I could walk places, surrounded by an abundant natural tropical landscape, where I can visit Kentucky friends and sit on their porches, I found myself choosing a neighborhood populated mainly by old-school white folks. Searching for my southern home, I looked for a place with a porch. Refurbishing a 1920s bungalow, similar to ones the old Sears and Roebuck catalog carried for less than seven hundred dollars with or without bathroom, I relished working on the porch. Speaking to neighbors who did not speak back, or one who let me know that they came to this side of town to be rid of lazy blacks, I was reminded how the black families who first bought homes in "white" neighborhoods during the civil rights era suffered — that their suffering along with the pain of their allies in struggle who worked for justice makes it possible for me to choose where I live. By comparison, what I and other black folk experience as we bring diversity into what has previously been a whites-only space is mere discomfort.

In their honor and in their memory, I speak a word of homage and praise for the valiant ones, who struggled and suffered so that I could and do live where I please, and I have made my porch a small everyday place of anti-racist resistance, a place where I practice the etiquette of civility. I and my two sisters, who live nearby, sit on the porch. We wave at all the passersby, mostly white, mostly folks who do not acknowledge our presence. Southern white women are the least willing to be civil, whether old or young. Here in the new

South three are many white women who long for the old days when they could count on being waited on by a black female at some point in their life, using the strength of their color to weigh her down. A black woman homeowner disrupts this racialized sexist fantasy. No matter how many white women turn their gaze away, we look, and by looking we claim our subjectivity. We speak, offering the southern hospitality, the civility, taught by our parents so that we would be responsible citizens. We speak to everyone.

Humorously, we call these small interventions yet another "Martin Luther King moment." Simply by being civil, by greeting, by "conversating," we are doing the anti-racist work of nonviolent integration. That includes speaking to and dialoguing with the few black folk we see from the porch who enter our neighborhood mainly as poorly paid, poorly treated workers. We offer them our solidarity in struggle. In King's famous essay "Loving Your Enemies," he reminded us that this reaching out in love is the only gesture of civility that can begin to lay the groundwork for true community. He offers the insight, "Love is the only force that can turn an enemy into a friend. We never get rid of an enemy by meeting hate with hate; we get rid of an enemy by getting rid of enmity. By its very nature, hate destroys and tears down; by its very nature, love creates and builds up. Love transforms with redemptive power." Inside my southern home, I can forge a world outside of the racist enmity. When I come out on my porch I become aware of race, of the hostile racist white gaze, and I can contrast it with the warm gaze of welcome and recognition from those individual white folks who also understand the etiquette of civility, of community building and peace making.

The "starlight bungalow" — my southern home for now, given the name assigned it in the blueprint of the Sears and Roebuck 1920s catalog (as a modern nomad I do not stay in one place) — has an expansive porch. Stucco over wood, the house has been reshaped to give it a Mediterranean flavor. Architecturally it is not a porch that invites a swing, a rocking chair, or even a bench. Covered with warm sand-colored Mexican tiles, it is a porch that is not made for true repose. Expansive, with rounded arches and columns, it does invite the soul to open wide, to enter the heart of the home, crossing a peaceful threshold.

Returning to the South, I longed for a porch for fellowship and late-night gatherings. However, just as I am true to my inner callings, I accept what I feel to be the architectural will of the porch and let it stand as it is, without added seats, with only one tin star as ornament. It is a porch for short sittings, a wide standing porch, for looking out and gazing down, a place for making contact — a place where one can be seen. In the old Sears and Roebuck catalog, houses were given names and the reader was told what type of life might be imagined in this dwelling. My "starlight bungalow" was described as "a place for distinct and unique living." When I first sat on the porch welcoming folk, before entering a dwelling full of light, I proclaimed, in old-South vernacular, "My soul is rested." A perfect porch is a place where the soul can rest.

In Kentucky my house on the hill has a long wide porch facing the lake that is our water source. This is not a porch for meeting and greeting. Perched high on a hill, the house and the porch has no passersby. Like the "starlight bungalow" this is a porch for "quiet and repose." It invites one to be still — to hear divine voices speak.

14

Aesthetic Inheritances: History Worked by Hand

To write this piece I have relied on fragments, bits and pieces of information found here and there. Sweet late-night calls to Mama to see if she "remembers when." Memories of old conversations coming back again and again, memories like reused fabric in a crazy quilt, contained and kept for the right moment. I have gathered and remembered. I wanted one day to record and document so that I would not participate in further erasure of the aesthetic legacy and artistic contributions of black women. This writing was inspired by the work of artist Faith Ringgold, who has always cherished and celebrated the artistic work of unknown and unheralded black women. Evoking this legacy in her work, she calls us to remember, to celebrate, to give praise.

Even though I have always longed to write about my grandmother's quiltmaking, I never found the words, the necessary language. At one time I dreamed of filming her quilting. She died. Nothing had been done to document the power and beauty of her work. Seeing Faith Ringgold's elaborate story quilts, which insist on naming, on documentation, on black women telling our story, I found words. When art museums highlight the artistic achievement of American quiltmakers, I mourn that my grandmother is not among those named and honored. Often representation at such shows suggests that white women were the only group truly dedicated to the art of quiltmaking. This is not so. Yet quilts by black women are portrayed as exceptions; usually there is only one. The card identifying the maker reads "anonymous black woman." Art historians focusing on quiltmaking have just begun to document traditions of black female quiltmakers, to name names, to state particulars.

My grandmother was a dedicated quiltmaker. That is the very first statement I want to make about Baba, Mama's mother, pronounced with the long "a" sound. Then I want to tell her name, Sarah Hooks Oldham, daughter of Bell Blair Hooks. They were both quiltmakers. I call their names in resistance, to oppose the erasure of black women — that historical mark of racist and sexist oppression. We have too often had no names, our history recorded without specificity, as though it's not important to know who — which one of us — the particulars. Baba was interested in particulars. Whenever we were "over home," as we called her house, she let us know "straight up" that upon entering we were to look at her, call her name, acknowledge her presence. Then once that was done we were to state our "particulars" — who we were and/or what we were about. We were to name ourselves — our history. This ritualistic naming was frightening. It felt as though this prolonged moment of greeting was an interrogation. To her it was a way we could learn ourselves, establish kinship and connection, the way we would know and acknowledge our ancestors. It was a process of gathering and remembering.

Baba did not read or write. She worked with her hands. She never called herself an artist. It was not one of her words. Even if she had known it, there might have been nothing in the sound or meaning to interest, to claim her wild imagination. Instead she would comment, "I know beauty when I see it." She was a dedicated

quiltmaker — gifted, skillful, playful in her art, making quilts for
more than seventy years, even after her "hands got tired" and her
eyesight was "quitting." It is hard to give up the work of a lifetime,
and yet she stopped making quilts in the years before her dying.
Almost ninety, she stopped quilting. Yet she continued to talk about
her work with any interested listener. Fascinated by the work of her
hands, I wanted to know more, and she was eager to teach and in-
struct, to show me how one comes to know beauty and give oneself
over to it. To her, quiltmaking was a spiritual process where one
learned surrender. It was a form of meditation where the self was
let go. This was the way she had learned to approach quiltmaking
from her mother. To her it was an art of stillness and concentration,
a work which renewed the spirit.

Fundamentally in Baba's mind quiltmaking was women's work,
an activity that gave harmony and balance to the psyche. According
to her, it was that aspect of a country-woman's work which enabled
her to cease attending to the needs of others and "come back to herself." It was indeed "rest for the mind." I learned these ideas from
her as a child inquiring about how and why she began to quilt; even
then her answer surprised me. Primarily she saw herself as a child of
the outdoors. Her passions were fishing, digging for worms, planting
vegetable and flower gardens, plowing, tending chickens, hunting.
She had, as she put it, "a renegade nature," wild and untamed. Today
in black vernacular speech we might say she was "out of control."
Bell Blair Hooks, her mother, chose quiltmaking as that exercise
that would give the young Sarah a quiet time, a space to calm down
and come back to herself. A serious quiltmaker, Bell Hooks shared
this skill with her daughter. She began by first talking about quiltmaking as a way of stillness, as a process by which a "woman learns
patience." These rural black women knew nothing of female pas-
sivity. Constantly active, they were workers — black women with
sharp tongues, strong arms, heavy hands, with too much labor and
too little time. There was always work to be done, space had to be

made for stillness, for quiet and concentration. Quilting was a way to "calm the heart" and "ease the mind."

From the nineteenth century until the present day, quiltmakers have, each in their own way, talked about quilting as meditative practice. Highlighting the connection between quilting and the search for inner peace, the editors of *Artists in Aprons: Folk Art by American Women* remind readers that:

> Quiltmaking, along with other needle arts, was often an outlet not only for creative energy but also for the release of a woman's pent-up frustrations. One writer observed that "a woman made utility quilts as fast as she could so her family wouldn't freeze, and she made them as beautiful as she could so her heart wouldn't break." Women's thoughts, feelings, their very lives were inextricably bound into the designs just as surely as the cloth layers were bound with thread.

In the household of her mother, Baba learned the aesthetics of quiltmaking. She learned it as meditative practice (not unlike the Japanese Tea Ceremony), learning to hold her arms, the needles — just so — learning the proper body posture, then learning how to make her work beautiful, pleasing to the mind and heart. These aesthetic considerations were as crucial as the material necessity that required poor rural black women to make quilts. Often in contemporary capitalist society, where "folk art" is an expensive commodity in the marketplace, many art historians, curators, and collectors still assume that the folk who created this work did not fully understand and appreciate its "aesthetic value." Yet the oral testimony of black women quiltmakers from the nineteenth century and early twentieth century, so rarely documented (yet our mothers did talk with their mothers' mothers and had a sense of how these women saw their labor), indicates keen awareness of aesthetic dimensions. Harriet Powers, one of the few black women quiltmakers whose

work is recognized by art historians, understood that her elaborate
appliquéd quilts were unique and exquisite. She understood that
folk who made their own quilts wanted to purchase her work because it was different and special. Economic hardship often compelled the selling of work, yet Powers did so reluctantly precisely
because she understood its value — not solely as regards skill,
time, and labor but as the unique expression of her imaginative
vision. Her story quilts with their inventive pictorial narrations
were a wonder to behold. Baba's sense of the aesthetic value of
quilting was taught to her by a mother who insisted that work be
redone if the sewing and the choice of a piece of fabric were not
"just right." She came into womanhood understanding and appreciating the way one's creative imagination could find expression
in quiltmaking.

The work of black women quiltmakers needs special feminist
critical commentary which considers the impact of race, sex, and
class. Many black women quilted despite oppressive economic and
social circumstances which often demanded exercising creative
imagination in ways radically different from those of white female
counterparts, especially women of privilege who had greater access
to material and time. Often black slave women quilted as part of
their labor in white households. The work of Mahulda Mize, a black
woman slave, is discussed in *Kentucky Quilts 1800–1900*. Her elaborate quilt "Princess Feathers with Oak Leave," made of silk and other
fine fibers, was completed in 1850 when she was eighteen. Preserved
by the white family who owned her labor, this work was passed
down from generation to generation. Much contemporary writing
on quiltmaking fails to discuss this art form from a standpoint which
considers the impact of race and class. Challenging conventional assumptions in her essay "Quilting: Out of the Scrapbag of History,"
Cynthia Redick suggests that the crazy quilt with its irregular design was not the initial and most common approach to quiltmaking,
asserting, "An expert seamstress would not have wasted her time
fitting together odd shapes." Redick continues, "The fad for crazy

quilts in the late nineteenth century was a time consuming pastime
for ladies of leisure." Feminist scholarly studies of black female experience as quiltmakers would require revision of Redick's assertions.
Given that black women slaves sewed quilts for white owners and
were allowed now and then to keep scraps, or as we learn from slave
narratives occasionally took them, they had access to creating only
one type of work for themselves — a crazy quilt.

Writing about Mahulda Mize's fancy quilt, white male art historian John Finley's comments on her work made reference to limitations imposed by race and class: "No doubt the quilt was made for
her owners, for a slave girl would not have had the money to buy
such fabrics. It also is not likely that she would have been granted
the leisure and the freedom to create such a thing for her own use."
Of course there are no recorded documents revealing whether or
not she was allowed to keep the fancy scraps. Yet, were that the
case, she could only have made from them a crazy quilt. It is possible
that black slave women were among the first, if not the first group
of females, to make crazy quilts, and that it later became a fad for
privileged white women.

Baba spent a lifetime making quilts, and the vast majority of her
early works were crazy quilts. When I was a young girl she did not
work outside her home, even though she at one time worked for
white people, cleaning their houses. For much of her life as a rural
black woman she controlled her own time, and quilting was part of
her daily work. Her quilts were made from reused scraps because she
had access to such material from the items given her by white folks
in place of wages, or from the worn clothes of her children. It was
only when her children were adults faring better economically that
she began to make quilts from patterns and from fabric that was not
reused scraps. Before then she created patterns from her imagination. My mother, Rosa Bell, remembers writing away for the first
quilt patterns. The place these quilts had in daily life was decorative. Utility quilts, crazy quilts were for constant everyday use. They

served as bed coverings and as padding under the soft cotton mattresses filled with feathers. During times of financial hardship which were prolonged and ongoing, quilts were made from scraps left over from dressmaking and then again after the dresses had been worn. Baba would show a quilt and point to the same fabric lighter in color to show a "fresh" scrap (one left over from initial dressmaking) from one that was being reused after a dress was no longer wearable.

When her sons went away to fight in wars, they sent their mother money to add rooms to her house. It is a testament to the seriousness of Baba's quiltmaking that one of the first rooms she added was a workplace, a space for sewing and quiltmaking. I have vivid memories of this room because it was so unusual. It was filled with baskets and sacks full of scraps, hatboxes, material pieced together that was lying on the backs of chairs. There was never really any place to sit in that room unless one first removed fabric. This workplace was constructed like any artist's studio, yet it would not be until I was a young woman and Baba was dead that I would enter a "real" artist's studio and see the connection. Before this workplace was built, quilting frames were set up in the spacious living room in front of the fire. In her workplace quilts were stored in chests and under mattresses. Quilts that were not for use, fancy quilts (which were placed at the foot of beds when company came), were stored in old-fashioned chests with beautiful twisted pieces of tobacco leaves that were used to keep insects away. Baba lived all her life in Kentucky — tobacco country. It was there and accessible. It had many uses.

Although she did not make story quilts, Baba believed that each quilt had its own narrative — a story that began from the moment she considered making a particular quilt. The story was rooted in the quilt's history, why it was made, why a particular pattern was chosen. In her collection there were the few quilts made for bringing into marriage. Baba talked often of making quilts as preparation for married life. After marriage most of her quilts were utility quilts, necessary bed covering. It was later in life, and in the age of modernity, that she focused on making quilts for creative pleasure. Initially she made fancy quilts by memorizing patterns seen in the houses of the white people she worked for. Later she bought patterns. Working through generations, her quiltmaking reflected both changes in the economic circumstances of rural black people and changes in the textile industry.

As fabric became more accessible, as grown children began to tire of clothing before it was truly worn, she found herself with a wide variety of material to work with, making quilts with particular motifs. There were "britches quilts" made from bought woolen men's pants, heavy quilts to be used in cold rooms without heat. There was a quilt made from silk neckties. Changes in clothing style also provided new material. Clothes which could not be made over into new styles would be used in the making of quilts. There was a quilt made from our grandfather's suits, which spanned many years of this seventy-year marriage. Significantly, Baba would show her quilts and tell their stories, giving the history (the concept behind the quilt) and the relation of chosen fabrics to individual lives. Although she never completed it, she began to piece a quilt of little stars from scraps of cotton dresses worn by her daughters. Together we would examine this work and she would tell me about the particulars, about what my mother and her sisters were doing when they wore a particular dress. She would describe clothing styles and choice of particular colors. To her mind these quilts were maps charting the course of our lives. They were history as life lived.

To share the story of a given quilt was central to Baba's creative self-expression, as family historian, storyteller, exhibiting the work of her hands. She was not particularly fond of crazy quilts because they were a reflection of work motivated by material necessity. She liked organized design and fancy quilts. They expressed a quiltmaker's seriousness. Her patterned quilts, "The Star of David," "The Tree of Life," were made for decorative purposes, to be displayed at family reunions. They indicated that quiltmaking was an expression of skill and artistry. These quilts were not to be used; they were to be admired. My favorite quilts were those for everyday use. I was especially fond of the work associated with my mother's girlhood. When given a choice of quilts, I selected one made of cotton dresses in cool deep pastels. Baba could not understand when I chose that pieced fabric of little stars made from my mother's and sister's cotton dresses over more fancy quilts. Yet those bits and pieces of Mama's life, held and contained there, remain precious to me.

In her comments on quiltmaking, Faith Ringgold has expressed fascination with that link between the creative artistry of quilts and their fundamental tie to daily life. The magic of quilts for her, as art and artifice, resides in that space where art and life come together. Emphasizing the usefulness of a quilt, she reminds us: "It covers people. It has the possibility of being a part of someone forever." Reading her words, I thought about the quilt I covered myself with in childhood and then again as a young woman. I re-membered Mama did not understand my need to take that "nasty, ragged" quilt all the way to college. Yet it was symbolic of my connection to rural black folk life — to home. This quilt is made of scraps. Though originally handsewn, it has been "gone over" (as Baba called it) on the sewing machine so that it would better endure prolonged everyday use. Sharing this quilt, the story I tell focuses on the legacy of commitment to one's "art" Baba gave me. Since my creative work is writing, I proudly point to ink stains on this quilt which mark my struggle to emerge as a disciplined writer. Growing up with five sisters, it was difficult to find private space; the bed was often my workplace. This quilt (which I intend to hold onto for the rest of my life) reminds me of who I am and where I have come from. Symbolically identifying a tradition of black female artistry, it challenges the notion that creative black women are rare exceptions. We are deeply, passionately connected to black women whose sense of aesthetics, whose commitment to ongoing creative work, inspires and sustains. We reclaim their his-tory, call their names, state their particulars, to gather and remem-ber, to share our inheritance.

15

Piecing It All Together

Watching quiltmakers do work by hand, I see in their labor an organic practice of mindfulness. Attention is concentrated, focused, repetitive. Sarah Oldham (Baba), Mama's mother, saw in the process of quiltmaking a way for a female to learn patience, the stillness of mind and heart that she would need as a grown woman to tend to work, home, and family. Learning to quilt in girlhood and continuing on into death and beyond, Baba was devoted to the ongoing practice of patience, combining spirituality with creative imagination. In stillness, sitting, sewing, she found herself able to listen more fully to the divine voice speaking, making god visible in the work. Baba was patient but she was not quiet. Creating beauty she found a way to speak, a way that moved beyond words.

Creativity is not quiet. I often experience the urge to create as a rumbling within the depths of my being. Like the tremors before an earthquake to come, that rumbling within me lets me know my senses have been aroused, stirred, that I can move into the imagination as though it is a fierce wave that will sweep me away, carry me to another plane, a place of ecstasy. The root meaning of the word ecstasy is to stand outside — that's what creativity does, it allows the creator to move beyond the self into a place of transcendent possibility — that place in the imagination where all is possible. And in that process one is both moved beyond measure and awed. In his insightful work *The Happiness Hypothesis*, author Jonathan Haidt writes: "Awe is the emotion of self-transcendence." It is precisely because artists recognize the vision that precedes the creation of a work emerges from a place we cannot locate or name, a place of mystery, that we stand before creation in awe. And this awe is not the province of those who are schooled or learned; it is democratic. It is an experience available to anyone irrespective of race, gender, nationality, class; it can be present to anyone who makes art.

Time and time again stories are told about the survival strategies folks use to maintain a sense of hope in desperate life-threatening situations of oppression, dehumanization, and violence. We tell stories about the ways we maintain a sense of worth and dignity in intolerable situations. I first heard such stories listening to Baba talk about living in slavery and beyond. She talked about hearing about the hardships, experiencing them as a girl, and she talked about the role of quiltmaking as both a functional necessity of life (making the covers that will keep out the cold, that will keep the body warm) but also she tells stories about the life-sustaining energy of the imagination, the artistry behind the creation of quilts.

In more recent times we can read the life stories of the black women quiltmakers of Gee's Bend, Alabama and be awed by their lives and work. From the location of newly acquired acclaim, they find a public voice to speak about the hardships they faced day to day, living as they say "a starvation life" where everybody was just struggling to get by, to make a way out of no way. And yet in a life that was more often than not filled with hardship, pain, and sorrow they found a pleasure of pleasure, of ecstasy, a place where they could transcend self, that place was artistic production — the making of quilts.

For years I held on to a pieced quilt top Baba had given me before I left home to go to college. The pattern was the Star of

David and each point of the star was cut from cloth made from summer dresses Mama and her sisters had loved and worn — worn so much they were worn out. But those soft cottons that in another culture might have become rags became a tapestry, a visual history of summer-time pleasure. Sitting in a rocking chair upstairs at 1200 Broad Street, Mama's childhood home, I would pore over this quilt as though it was indeed a text to be read and Baba would tell me the story of each dress and the girl who wore it. I could imagine then Mama and her sisters, beautiful young girls, delighting in summer, wearing their much wanted, much beloved favorite dresses.

I took this pieced quilt top on every one of my life's journeys, packing it in soft cotton pillowcase covers ('cause a quilt should never be placed in plastic). There was no place I lived where I did not find those special moments to share this precious treasure with anyone with an eye for beautiful old things. But whenever I was asked to give the quilt top over to some helping hand that would be willing to quilt it for me, I could not trust enough to let it go. Afraid it might be lost, as my Baba was lost, gone to a place where she will sew no more. She did not imagine herself doing much in the afterlife that she had done in her life before. Heaven to her was one big porch where she was going to just sit and sit and enjoy the warmth and cool sunshine. I suppose the sunshine could not be hot because that might suggest she was in a hell somewhere. No, she was up there on the heavenly porch just sitting and looking at the angels go by. Just as she sat on her porch in Kentucky and watched the world. When we were young'uns she was never on that porch alone but as she grew old and, as the preacher said when she passed away, "her choicest friends were gone"; she would sit on the porch rocking all by her lonesome, just looking out at the world.

I wonder how she sees me now. I wonder what she thinks of me coming home to live in Kentucky. No doubt she is certain she was right in thinking it unnatural to live away from your people and that to save your soul the best you can do is come home, even if you are coming home to die. We saw a lot of that when we were growing up. Folk who had lived in northern cities all their lives coming home to die. Surely Baba knows I have come home to live and, yes, to one day die, but hopefully no time soon.

Here, the place I have chosen to make home is known for the presence of Kentucky artisans, many of them quiltmakers. Showing my grandmother's quilt top to neighbor and friend Alina Strand, I hear from her that she knows just the right somebody to quilt this piece for me. Listening to me share my fears, about losing this last living piece of intimate childhood shared with Baba, she assures me she understands. But she says with her blunt, I-know-what-I-am-talking-about voice: "Give me this quilt top to give to Miss Pauline. She will know what to do with it and she will do it right." This is a mountain woman, she tells me, "who has been quilting since she was a girl. She will sew your quilt by hand. You'll see." I wait until my sister V. is visiting before I will hand over the quilt top. I need there to be a witness. I tell Alina: "What if I were to die and you all just forgot all about my quilt?"

When I am gone, I tell them, "this quilt has got to stay in the family somewhere. It has got to tell the story." That's what the older Gee's Bend quiltmakers share, that a quilt is as much a document tracing the story of lives as it a comforting cover. In her introduction to the big book about their work, *The Quilts of Gees Bend*, Alvia Wardlaw tells of the cultural geography documented in this ·collective story of quiltmaking: "The women of Gees' Bend regard their quilts not only as gifts for family, but also emblems of their own unique natures — a combination of industry and ingenuity, a singular hallmark of their capabilities as homemakers." In all ways Baba took pride in her home, from the eggs her chickens laid, the flowers in the yard, the fresh fish caught at the creek, or the quilts she made through times. It was all testimony of her artistry — her skill at self-invention, her power to be self-determining.

The day I spread out Baba's "Star of David" quilt was just a day of happiness. It touched my soul this meeting of two grand quilters, one dead, one living, but both knowing how to breathe life into scraps of cloth. When Miss Pauline finished the quilt it was a beautiful work to behold; it brought tears to our eyes. The quilting sewn by hand was itself a work of art. It was as though each point of the star was sewn with a tender watchful caring eye. The quilting called for a conversation between the dead and the living. Like many older white women living in the Kentucky hills, Miss Pauline had little contact with black folks growing up, but with this quilt she was making a connection. She and her sisters formed a bond with Sarah Oldham, learning her from the way she sewed, from the way she put the pieces together. Miss Pauline laughed as she shared with me the details of Baba's character that she had gleaned from a pieced quilt top telling me "she was a strong woman — a woman who knew her own mind." Geraldine Westbrook, one of the Gee's Been quilters who started young, testifies: "I don't follow no pattern. ...When you sit down you got to get yourself a mind of your own, figure out a way to put them together." Feminist thinkers have only begun to look at quilting, whether individually or collectively, as conscious-raising work that is rarely created by a subjugated being. Alvia Wardlaw affirms this when she writes about the Gee's Bend quilters creating public art, declaring: "Then she has made her own statement of bold independence, almost defiant in a sense, because in the face of such near tragic epic sagas of poverty and misery, she has had ·the audacity to create something bright and beautiful that has never been seen before and will never be seen quite that way again and it is all hers ..." Clearly this is the power of imagination, that it can transform us, that it can spark a spirit of transcendent survival.

For the spirit of self-reliance and self-determination that was aroused and is aroused by quiltmaking, by this fusion of the practical with the artistic, stirs the imagination in ways that almost always lead to emotional awareness and emotional growth. That spirit of self-reliance often creates the social context that made survival possible, that made it is possible for there to be moments of triumph and possibility. Even though most of the older Gee's Bend quilters talk about growing up hard, facing adversity, a great majority of them have lived a long time. And they lived into their glory. They were all hard workers who labored in the fields as consistently as they pieced quilts.

Certainly in times past most homemakers did not separate hard outdoor work on the land from the work done inside domestic households. There is much to be written about the connection between ecopsychology and the art of quilting. Almost all the elder black women who made awesome quilts that I knew growing up also worked growing gardens — food and flowers. Indeed, the quilt on my bed today is called "grandmama's flower garden." In the history of black women quiltmakers we see an empowering imagination revealed.

Writing about the power of the imagination in an essay on the civil war, Wendell Berry emphasizes that "the particularizing force of imagination is a force of justice. ... Imagination, amply living in a place, brings what we want and what we have ever closer to being the same. ... If imagination is to have a real worth, to us, it needs to have a practical, an economic effect. It needs to establish us in our places with a practical respect for what is there besides ourselves. I think the highest earthly result of imagination is local adaptation. If we could learn to belong fully and truly where we live, then we would all finally be native Americans, and we would have an authentic multiculturalism." No one spends much time talking about quilts and justice, about quilts and the uses of the imagination. And yet the local functional use of quilts created to keep a family a warm was a major gesture of recycling, of letting nothing go to waste. This economic use of cloth was and is an important contribution to the development of sustainability. Quilts are an amazingly democratic art form. They have been created and used by folks of all classes and races. They are a grand symbol of democracy. They keep folks warm while at the same time bringing beauty and creativity into their lives. As a form of public art quilts have played a major role in projecting visions of aesthetic beauty into the lives of the poor, of those who lack economic power and privilege, who may or may not be educated, who may have no book learning.

Baba could not read or write. But you would only have known this fact of her life (it was an aspect of her childhood she spoke of with bitterness in her memory for she was forced to leave school at an early age to work in the fields), if she shared that information. The interior landscape of her world was one of constant creative engagement. She was making something. And like most Kentucky backwoods folk she could talk and talk and tell a right good story. Her stories that live are there in her quilts. Were they all collected together and shown they would reveal a culture of place carefully, imaginatively constructed.

Fortunately, the Gee's Bend quiltmakers have received national attention and as a consequence there may be greater awareness of the artistry of quiltmaking in the lives of poor and working-class southern black women. It is vital that attention to this culture of place not be seen as a dead or dying expression of creativity, to be treasured because it is no more. Instead it should be regarded as part of a continuum where this legacy of creativity, of using the imagination to enhance well-being, if fully honored will no doubt take new and different forms. There are individual black women quiltmakers whose work will never gain public recognition. This does not diminish its transformative power in its own isolated local culture. For whether present in a mansion or a shack, the quilt can open our eyes to beauty, to an aesthetics of possibility that is infinite.

16

On Being a Kentucky Writer

Being a Kentucky writer is for me a question of upbringing and sensibility. All the writing I have done and currently do has the particular flavor of my growing up in rural Kentucky hills as a child, then later in town. This sensibility can be likened to a seasoning in food preparation. A particular herb or spice used to flavor a dish may make it distinctly different — unusual, outstanding even. Those who eat the dish may taste the difference with no clue as to where the difference is coming from. This same dish may be cooked and eaten all over the world, but prepared by diverse cooks it may have distinctive flavors that make it stand out. Like many great and good cooks, I might know what flavor makes my dish unique but never tell. In the telling it may lose it delicious seductive essence.

I begin this discussion of being a Kentucky writer with the analogy with food in part because the home cooking of Rosa Bell, my mother, who is a great cook (even in her old age as she struggles with the loss of short-term memory), nourished my body and spirit. When I first left Kentucky to live elsewhere and could not find in new places food like that I had intimately known, I felt utterly strange. Eating food foreign to my tastes and appetite was a constant reminder of the distinctiveness of the world I was coming from. The food I ate growing up was not known away from my home. I ate food no one in my world away from Kentucky was accustomed to eating.

Those appetites, those familiar longings so inviting to my senses, could not be easily shared. I stored them. I put them away. I felt no need to talk about my hunger for familiar food. I felt no need to explain the sense of profound lack haunting me at meal time. On those rare occasions where I felt the need to explain, words were never fresh or tasty enough. They could not convey a sensibility learned mouth to mouth, heart to heat. Even now when I go to the family home, Rosa Bell cooks the familiar foods — nurturing a particular body and soul, recipes handed down by generations of Kentucky women.

Geography shaped my perspective in ways that made it more complex. Coming from a rural Kentucky small town into the wide world beyond distance clarified for me the uniqueness of my bluegrass sensibility. Unlike most of my college peers I had been compelled by circumstances of birth and origins to face the intersections of geography, race, class, gender. When I left Hopkinsville, Kentucky, to attend college, geography more so than any other factor shaped my destiny. My Kentucky accent always separated me from peers. And even though it did not take long for me to change the way I talked, in my dreams I spoke in the vernacular of homeplace. Learning to be silent about the ways of thinking and being I had learned as a Kentucky girl, a child of the backwoods did not mean that this sensibility did not continue to be the foundation of my thoughts and actions.

Writing was the place where I could best express that sensibility without overtly calling attention to geography. Since I had been taught as an English major both in undergraduate and graduate classes that the goal of a writer was to be universal, I endeavored to keep the "I" and the personal self out of writing, or at least buried in the writing so deep it would be difficult to find. This equation of universal with the absence of personality in writing I came to realize is a false dichotomy, for it is in calling out our specific and unique sense of who we are in that we invite that empathic identification which makes the specific also encompass the universal. During those years when I was seeking to gain affirmation from an academic elitist white (usually) male hierarchy, I allowed my self and my personal voice to be diminished by a symbolic cultural imperialism that used a false focus on being universal to mask the loud and aggressive sound of the particular world view of elitist white males.

In his work of fiction, *The Language of Cannibals*, George Chesbro critiques dominator culture, stating that "the fastest way to destroy a society is to corrupt language." Indeed, he maintains "lies are the language of cannibals." The dedication to truth that has been a driving force in my life and work was a value instilled in me growing up. It was an essential part of the cultural ethos of the Kentucky backwoods, of the hillbilly country folk who were my ancestors and kin. Truth was central to their resisting anarchist mindset, their rebellion against established norms. They prided themselves on their ability to cut through the false and fake to find the real authentic treasure. Rebellion against false authority was essential to the freedom my backwoods kin taught me.

Yet these beliefs merely made me feel as though I was a stranger in a strange land when I left the poor and working-class world of my growing up in Kentucky to attend an elite college. There I found that folks took my concern with integrity and honesty to be naïve, just another habit of being that marked me as county and unsophisticated. Like all Kentucky folks who embraced the anarchist values of the backwoods, I considered myself an outsider even before I left my homeplace. Writing about the spiritual significance of outsiderness, Judy Lief offers her understanding of what it means to stand outside: "The outsider is ready to speak out when others are silent, ready to challenge conventional wisdom, ready to sacrifice her own comfort and reputation in the service of turning people from despair and reconnecting them with what is sacred. The outsider, through personal example, presents an alternative vision of reality, an alternative way of living your life." While I understood how to live as an outsider in the Kentucky hills, I did not know how to survive and thrive as an outsider in a more conventional world.

Like many outsiders engaged in the arts I turned to the world of bohemian culture, of writers and artists to receive a new life map, a blueprint for how to be alternative hip, cool. That world was no more affirming of geographical outsiderness than the conservative world of conservative biases and hierarchies. Being country folk, having a "country" sensibility, was as uncool in hip circles as it was in mainstream culture. Significantly, the world of hedonistic cool was as disdainful of commitment to truth, honesty, and a life of integrity as the culture of cannibals. When as poet and visual artist I found myself firmly ensconced in the world of cool, I felt the need to keep all those Kentucky based values hidden deep in my own personal and private geographical sub-culture.

The feminist movement with its focus on recovering women's history and telling women's stories was the social and political context where the demand that women reclaim our individual and collective voice was linked to an appreciation for difference, for vernacular culture. It was impossible to conjure the voices of female ancestors and not hear in their stories, their words, the nuances of the Kentucky hills, the strong accents of folk speech, the hidden roots of old English and Appalachian cadences. To return to the voice of the primal mother, I had to return to my own vernacular Kentucky speech. The voice I never used in the college environment was the voice that was needed to reaffirm my connections to homeplace.

Kentucky as the homeplace of my mind and heart is both real and mythic, distinct from the concrete experience of living in the bluegrass state. The Kentucky I conjured in geographical exile was always sweeter than the real-life culture of the backwoods. When I left home I carried deep and profound memories of old-time religion and hardcore church-going, but I did not seek out church homes in my new locations. Identifying myself with the backwoods outlaws of my growing up who eschewed any notion that the holy spirit and spirituality could be institutionalized, I claimed my spiritual practice in private away from the conventional gaze of patriarchal religion and father god.

As my writing career developed and I became more well-known when interviewed, no one asked if I saw myself as a Kentucky writer. Writing primarily non-fiction, usually social and political critique that did not evoke a geography of place, my work was rarely seen as having a vital connection to the region of my upbringing. When I begin to write cultural criticism often using personal experience as a framework, I wrote openly about my Kentucky past but rarely did I directly identify my home state. Readers simply saw and see me as writing about "southern" roots. Had my writing been fiction, perhaps, there might have been more critical acknowledgment of the role geographical location as place of origin and object constancy played in the cultivation of my sensibility as a writer.

The more well-known one becomes as a thinker, as a writer, particularly if you did not begin life as a member of any privileged group, the more you are asked to explain how you became the writer you are. To me, all the years of my life growing to young womanhood in Kentucky and the years coming home represent to me the foundation of all that I have become as a critic and a writer. All the eccentric sensibility of the Kentucky backwoods, its nuances and particular flavors mixed with all the other experiences I have had, make me all that I am.

Like poke, a Kentucky favorite which changes the flavor when added to turnip, collard, or mustard greens, the hillbilly culture, the backwoods ethos is that particular ingredient which shapes and forms me. It is that foundation that leads me to embrace wholeheartedly the reality that I am indeed a Kentucky writer.

17

Returning to the Wound

When I began teaching at Berea College, the first short seminar focused on the work of Kentucky writer Wendell Berry. Reading and writing poetry first led me to his work. Excited to discover in my late teens a Kentucky writer whose work bespoke an interior landscape that I understood intimately, I embraced his work wholeheartedly. To read Berry writing such powerful lines as "we hurt and are hurt and have each other for healing" was to enter that space where words renew the spirit and make it possible for one to hold onto life. Once I discovered the work of Wendell Berry, I read everything he had written that I could find.

His vision of a culture of place where one makes a homeplace in a world rooted in respect for all life, earth and community, where there is spiritual grounding and aesthetic celebration of beauty, where there is a pure enjoyment of simple pleasures, was for me a guiding light. My development as intellectual, critical thinker, poet, essayist, "writer" has followed a path charted by Berry. Like Berry, I write poetry, essays, fiction, and cultural criticism. Even our movements out into the world have had a similar trajectory writing and teaching at Stanford University in California, working in New York City, ultimately returning to Kentucky — to make home forever in our native place. Twenty years separate our experiences. When Wendell returned to Kentucky and bought a small farm in the spring of 1964, I was still a typical Kentuckian, I had never been away from my native place and there was no thought in my mind then that one could live in any other place.

Reflecting on his return Wendell declares: "That return made me finally an exile from the ornamental Europeanism that still passes for culture with most Americans. What I had done caused my mind to be thrown back forcibly upon its sources: my home countryside, my own people and history ... It occurred to me that there was another measure for my life than the amount or even the quality of the writing I did; a man, I thought, must be judged by how willingly and meaningfully he can be present where he is, by how fully he can make himself at home in his part of the world. I began to want desperately to learn to belong to my place." My sense of belonging in a culture of place has been profoundly shaped by the words and wisdom of Wendell Berry. He is for me and many other readers a prophetic voice. When asked in a interview whether he was comfortable with being seen in this way, he responded, "we all ought to be prophets in the sense that we should see the truth and tell it."

One of the most profound truths of our nation is the entrenched racism and white supremacy that continues to inform the politics of daily life. When civil rights initiatives failed to create a program for the systematic destructions of racism both in theory and practice, many citizens of our nation just began to feel that race and racism were a topic that should simply not be openly discussed. This was not the case for Wendell Berry. Indeed, in response to the civil rights struggle to end racism, he wrote *The Hidden Wound*. This is the work by Berry that I consistently teach. There is little public discourse on the subject of race and racism in the state of Kentucky. Shame-based

memory of both past and present domination and subjugation of
black people by white people has led to a deep silence which must be
continually broken if we are to ever create here in our native place
a world where racism does not wound and mark all of us every day.

The Hidden Wound is both memoir and critical reflection on the
state of race and racism. Berry contends: "It seems to me that racism
could not possibly have made merely a mechanical division between
the two races; at least in America it did not. It involves an emotional
dynamics that has disordered the heart of the society as a whole
and of every person in the society. It has made divisions not only
between white people and black people, but between black men
and black women, white men and white women; it has come between white people and their work, and between white people and
their land. It has fragmented both our society and our minds." To
counter this fragmentation we must confront one and consistently
challenge each other to do the work of ending racial domination,
of ending racism within and without. Writing The Hidden Wound
was a way for Berry to take part in the civil rights struggle by both
offering a more complex understanding of social relationships between black and white people formed even in the midst of intense
racial apartheid.

This work is important testimony. Long before contemporary
cultural studies made the study of "whiteness" a crucial discipline
necessary for any complete understanding of the way in which racism has shaped our national consciousness, Berry was engaged in
critical thinking about just this subject. In The Hidden Wound he presented a critical reading of the way in which whiteness constructed
as an identity rooted in domination — the need for an exploited and
oppressed other — served to distort reality for whites and blacks
alike. Importantly, Berry begin his deconstruction of whiteness by
examining his family history, the relationships between himself
and the black folks that were a part of his childhood life in Henry
County, Kentucky. He paints an intimate portrait of his relationship

with two black folks who worked and lived as part of the extended community on his family land. In his remembering of social interaction with Nick Watkins and Aunt Georgie, Berry endeavors to show that dominator culture and the racial apartheid it upheld could not prevent intimacy from emerging between black and white folk. And he emphasizes that such intimacy always humanizes, even though it forms itself within a dehumanizing social framework.

Throughout his description of his childhood relations with Nick and Aunt Georgie, Berry is careful to acknowledge that white privilege may be informing his perspective, that he may be describing these two folks in ways that they would not have described themselves. And while there is no doubt that his personal reflections include at times sentimental accounts of a boyhood bond with two black folks, his intent is always to make clear to the reader their essential humanity. He wants readers to understand that despite the persistence of racism, these inter-racial social relations rooted in mutual acknowledgment, that there was more to life than race that made it possible for individual black and white folks to love one another. With the lives of these two seemingly subjugated ordinary black people as backdrop, he attempts to show readers a world of emotional intelligence and awareness powerful enough to mediate domination and make equality of longing and desire the measure of meaning and not skin color. He states: "And so I have not written at such length out of my memory of Nick and Aunt Georgie in order ... in order to reexamine and to clarify what I know to be a moral resource, a part of the vital and formative legacy of my childhood. The memory of them has been one of the persistent forces in the growth of my mind. If I have struggled against the racism that I have found in myself, it has been largely because I have remembered my old sense of allegiance to them. That I have gone back to my native place, to live there mindful of its nature and its possibilities, is partly because of certain things I learned from them of that they exemplified to me." Significantly, the portraits Berry paints of Nick

Watkins and Aunt Georgie are evocative of folks I grew up with in the segregated world of my childhood.

Berry was one of the first Kentucky writers to document in non-fiction the special oppositional consciousness of subjugated black people that I often write about. In the many essays I have written about family and community, I describe a culture of belonging where folks like Nick and Aunt Georgie are the norm. Berry acknowledges that they both had a profound engagement with the natural world, with sustainability, with a metaphysical universe that is beyond race. Indeed, he intends readers to understand that despite the power of racism, to a grave extent, Nick and Aunt Georgie lived an interior life of their own invention. The values and beliefs governing the world they made for themselves shaped their actions as much if not more than the imposed constraints of racial domination. As I began to remember and write about the unique black elders who had shaped my life vision, I began to see the importance of Wendell's documenting the lives of Nick Watkins and Aunt Georgie.

Since history has denied so many poor and disenfranchised citizens of our nation a voice, no matter their color, all contributions that document and give voice to diverse experiences are needed. I needed to place the stories of Nick and Aunt Georgie alongside my stories of Baba, Daddy Gus, Sister Ray, Daddy Jerry, and so many other black folks who farmed Kentucky land, who taught us about the importance of nature, of listening and believing in divine spirit. Our shared blackness does not mean that the stories I tell of their lives would be what they would speak if their voices were doing the telling. But it is vital that their visions — that their will to live lives informed by transcendent notions of freedom and possibility be on record. I would not be all that I am today without their witness. Wendell echoes this sentiment when he speaks of Nick and Aunt Georgie.

Significantly, as I have traveled around our nation asking folks what is the force that has mostly led them to resist domination culture, to stand against racist domination and oppression, the response is almost always the presence of love. In my critical writing on ending racism, I have talked about the role of choice, of loving justice and making a commitment based on that love. Importantly, in speaking against any notion that racism cannot be changed, that it represents some cultural absolute, I have called attention to white children who early on in life refused to accept the mantle of white privilege and embrace racism. One crucial aspect of *The Hidden Wound* is Berry's sharing of his own conviction in childhood that there was no need to separate people on the basis of race. As a child he understood and lived within the politics of racial segregation. And it was in that childhood that he learned to resist.

I am often asked by committed white folks who sincerely want to see racism end to provide them a map of what they might do. My response is always to share that it is they who must bring critical awareness to the places in which they live and discover there what it is they must do. Wendell narrates just such a moment, of refusing white privilege, and taking a stand for love and justice in *The Hidden Wound*. His grandmother plans a birthday party for her grandson but she does not invite Nick as the social mores of her time governed by racism would have frowned upon black and white folks socializing together formally. When the boy invites his beloved friend Nick to the party, the unease that is stirred up around him creates the awareness that he has "scratched the wound of racism" and that everyone around him feels the pain. He recalls: "It was suddenly evident to me that Nick neither would nor could come into the house and be a member of the party … If Nick has no place at my party, then I would have no place there either; my place would be where he was." By choosing to give up white privilege, the boy was able to create a zone of mutuality beyond race where "we transcended our appointed roles." This small act of resistance wherein the boy refused to stand with racism but rather marks a path of resistance wherein he stands outside is the true meaning of solidarity. And it is the enactment of such solidarity that is racism's undoing. No doubt that is why Wendell can testify that at that moment: "I was full of a sense of loyalty and love that clarified me to myself as nothing ever had before." When teaching *The Hidden Wound*, I ask students to think about how this simple story, the gesture of friendship, works to humanize both individuals whom the enacted practice of racism would disfigure and distort. To end racism white folks who have accepted unearned white privilege must be willing to forego those rewards and stand down, expressing their solidarity with those to are the most immediate victims of racist assault and domination.

In *The Hidden Wound*, first published in 1968, Wendell Berry was prescient in his insightful critique of whiteness, showing himself to be among the first well-known cultural critic to see and publicly name the link between white racist domination and destruction of the earth. He does not sugarcoat his critique boldly proclaiming: "the white race in America has marketed and destroyed more of the fertility of the earth in less time than any other race that has ever lived. In my part of the country, at least, this is largely to be accounted for by the racial division of the experience of the landscape. The white man, preoccupied with the abstractions of the economic exploitation and ownership of the land, necessarily has lived on the country as a destructive force, an ecological catastrophe, because he assigned the hard labor, and in that the possibility of intimate knowledge of the land, to a people he considered racially inferior; in thus debasing labor, he destroyed the possibility of a meaningful contact with the earth." Berry acknowledged that agrarian subjugated black folk were able to work the land and "develop resources of character and religion and art that have some resemblance to the old world." Displaced African people found working the land to be one of the few locations where ties to their landscape of origin could be reclaimed.

In seeking freedom in the city via mass migration from the agrarian South, most black people began to embrace dominator ways of thinking about the earth. Berry contends: "The move from country to city, moreover deprives them of their competence in doing for themselves. It is no exaggeration to say that, in the country, most blacks were skilled in the arts-of-make-do and subsistence ... They knew how to grow and harvest and prepare food. They knew how to gather wild fruits, nuts, and herbs. They knew how to hunt and fish ... In the cities, all of this know-how was suddenly of no value ... In the country, despite the limits placed upon them by segregation and poverty, they possessed a certain freedom in their ability to do things, but once they were in the city freedom was inescapably associated with the ability to buy things." Of course, not all black folk migrated to cities. And it is the memory of a sustained oppositional living sub-culture, like the one Berry describes, that offers a glimmer of hope in the present day. Hence the importance of both naming black folks' collective estrangement from our agrarian past and taking steps to uncover the true nature of the culture of belonging as well as the naming of the trauma that took place when country life lost meaning and visibility.

This estrangement from our agrarian past, this rupture, can only be healed by full acknowledgment of that legacy and the functional use of that legacy in the present. Remembering Nick Watkins and Aunt Georgie (and folks like them) is one way to intervene on our nation's collective forgetting. One of the silent spaces in Berry's narrative is caused by his lack of familiarity with the more developed and articulated land stewardship of Kentucky black people. He learns some of that sub-culture of blackness from his conversations with southern black writer Ernest Gaines. In Berry's short story "Freedom," a fictionalized account of Nick's funeral, he shares accurate secondhand knowledge of the unique way many southern black folks approach death.

In *The Hidden Wound* Berry certainly shows both a keen awareness and a profound respect for the humanizing culture black folks created in the midst of adversity. Rightly, in the afterword added in 1988, twenty years after the initial publication, he still acknowledges "the freedom and prosperity of the people" cannot be seen as separate from the issue of the health of the land" and that "the psychic wounds of racism had resulted inevitably in wounds in the land, the country itself." While Berry can state that he believed then and now "that the root of our racial problem in American is not racism" but "our inordinate desire to be superior." Of course were Berry a student of my work, I would encourage him to think more about the ideology of white supremacy which has proven to impact the psyches of white dominators of all classes as well as people of color, especially black people so those of us who have not decolonized our minds collude in perpetuating the very structures of racist domination. It has been far easier for our society to rid itself of many of the outward mani-festations of racial inequality than is has been to rid this nation of the western metaphysical dualism that gives white supremacist thinking its foundation.

As Berry foresaw, separation of the races into work and living environments where there is no chance that humanizing encounters will occur, interactions that would challenge biases and stereotypes, has simply led to an intensification of white supremacist thought and action. Recent public discussion of race and racism show that this notion has advanced little in the direction of ending racism. And it has become more than evident that legislation that has created greater racial equity has done little to change the nature of social relations between diverse groups of people. What has become clear is that education for critical consciousness coupled with anti-racist activism that works to change all our thinking so that we construct identity and community on the basis of openness, shared struggle, and inclusive working together offers us the continued possibility of eradicating racism. The struggle to end racial discrimination and domination must be renewed. And one of the clear means of challenging and changing racism is the will to change our interior lives, to live differently. *The Hidden Wound* continues to be a useful discussion of importance to think about race, especially when it calls all of us to think of people of color, black people, as more than simply victims. It has been so detrimental for black people to internalize the notion that we are always and only victims. There is much to be gained by studying the lives of elder black people who mapped an oppositional world view, a culture of belonging that was both humanizing and spiritually uplifting, one that helped us understand that we are always more than our pain,

One of the most insightful ideas presented in *The Hidden Wound* is the acknowledgment that inter-racial living, even in flawed structures of racial hierarchy, produces a concrete reality base of knowing and potential community that will simply be there when all that white and black folks know of one another is what they find in the media, which is usually a set of stereotypical representations of both races. The love that flourished in Berry's heart for Nick, a love so deep it would lead him to write in another fiction based on this relationship, "if he loved me as I loved him, I was indeed blessed." We should all be so blessed as to engage in social relations that are not tainted and distorted by the twisted politics of racist thought and action.

Like Wendell Berry, I believe that we can restore our hope in a world that transcends race by building communities where self-esteem comes not from feeling superior to any group but from one's relationship to the land, to the people, to the place, wherever that may be. When we create beloved community, environments that are anti-racist and inclusive, it need not matter whether those spaces are diverse. What matters is that should difference enter the world of beloved community, it can find a place of welcome, a place to belong.

18

Healing Talk:
A Conversation

On both days that I journeyed to Wendell Berry's home in Port Royal, Kentucky, with my friend Timi Reedy (sustainability and holistic life consultant), the sun was shining brightly and the skies were that brilliant shade of breathtaking blue. As we traveled I began to hear a song in my head that we sang in grade school: "I look at the sky, the clouds floating by, the blue like no blue on earth could be. I greeted the air and whispered a prayer, for god made this loveliness for me." And as the song in my head ended, I repeated aloud a line of poetry "the world is charged with the grandeur of god." Those two journeys were heavenly. I was on my way to see and talk with Wendell Berry poet, essayist, novelist, cultural critics, farmer, a fellow Kentuckian whose work has influenced my intellectual development. I was downright giddy. Our first visit there we sat with Wendell and Tanya (his wife of long years) around their kitchen table eating pound cake, drinking tea, and talking. There was much wit, laughter, and general silliness on display as well as much serious and deep thought. There was no tape recorder documenting our conversation. In the past I have often scoffed at those folks who cannot go anywhere without a camera, a recording device, video, without some instrument to document for the future. Now that I have witnessed the deep pain and grief that can be caused by loss of memory, through illness, dementia and Alzheimer's (my mother begin to fall into a world of profound forgetfulness shortly before my first visit to the Berry household), I can acknowledge the value of documentation for a future time. I know firsthand what a blessing it is to have a record — a way to remember that goes beyond the mind.

The second time I go to Port Royal, there are only three of us present: Wendell, Timi, and myself. This time we sit on the porch, talking, recording. Our conversation is not as magical as it was that first time. We are more serious. We talk together across our differences of age, race, sex. Almost twenty years separate us in time. As a black girl growing up in the Kentucky hills and later in a small segregated part of my town, my experiences were concretely different from that of Wendell's growing up white, male, privileged. Yet, here we are, two Kentucky writers, one old and one not so old, finding a place of closeness despite all that would and could separate us. We do not think alike about all matters but there is so much shared understanding that our words seem to belong together as we talk in our native place. Although I could have talked with Wendell endlessly about poetry, writing, farming, beauty, I chose to speak with him about race and racism, topics that more often than not are not addressed when folks interview Wendell. It was my hope that our words would break through the profound racial silence that is present in public discourse on the subject, a silence that must be broken if we are to truly find ways to end racism.

Meeting with Wendell Berry, talking and laughing with him, I am grateful that we can transgress the boundaries that would keep us from talking together, from knowing one another, boundaries of race, class, experience. Talking together, we listen and hear beneath our words the possibility of making beloved community. We experience that movement beyond race where we can talk together and let our hearts speak.

BH: Wendell, you were one of the first thinkers to insist that we cannot have health in mind, body, and spirit if we don't have health in relationship to the land. You are one of the few elder white male writers who has consistently linked the persistence of racism to destruction of land and people reminding us that we cannot have a world of great environmentalism while we maintain the violence of white supremacy and racism.

WB: I hope we are more than a few. Your summary of the general position is about right. We are under obligation to take care of everything and you can't be selective if you are going to take care of everything,

BH: Your book, *The Hidden Wound*, continues to be a useful contribution to discussions of race and racism in our nation. It is always particularly relevant to me because you are one of the few writers of any race who understands the agrarian history of African-American folk in this nation.

WB: A writer who really understands this subject is the African-American writer Ernest Gaines.

BH: Absolutely. I have been writing about his work referring to the comments you make in the essay "American Imagination and The Civil War" where speaking about Gaines you explain to readers: "He has imagined also the community of his people as part of the life of their place and the hardships of that community. He has imagined the community's belonging to its place, the houses that had the names of people, the flower-planted dooryards, the church, the graveyard, the shared history and experience, the shared stories, the talk of old people on the galleries in the summer evenings and the young people listening. He has also imagined the loss of these things. ... He has shown that the local, fully imagined, becomes universal." He writes so poignantly about that loss when he describes in *A Gathering of Old Men* a scene where white folks talk about plowing over the graves of black folks as a metaphor for erasing the history of the black farmer, erasing even evidence that black folks were ever stewards of the land.

WB: That they ever had an intimate relationship. Did you know that Ernie has bought land down there where he grew up. And he's bought the church, the building that serves his church and school and moved it to this property.

BH: Yes, he's been a role model for me and other black folks returning to the South from the North. Like me he was probably influenced by Wendell Berry. I know it was following your lead that influenced me to buy my Kentucky land, to assume a stewardship position. In my case it's just about keeping green space, protecting the land from development. I'm not a farmer. I don't have a green thumb, but I know I have fifteen acres of land that will be "forever green."

WB: Well, now you may have to look at the possibility that I was influenced by Ernie. You know Ernie and I started out together.

BH: Tell me more. I didn't know this.

WB: We were both Stegner Fellows in the writing program at Stanford in 1958. There we were, two young men, one white and one black, with all this common knowledge. We both knew things that nobody else in our class knew. Then there came a time when Ernie and I had to have a little talk and lay it all out between ourselves. You know you don't understand everything at the start — you grow into understanding. The two people I grew to understand in that seminar were Stegner himself and Ernie.

BH: Do you still believe, Wendell, as you wrote in 1968, that the lives of white people are conditioned by the lives of black folks, that there is an ongoing "emotional dependency."

WB: Well, if you reduce any group of people to a set of stereotypes,
you impair your mind and that was what was going on at Stanford in 1968. The university had invited some black students to
enroll. It was not long after the Watts riots. The black students
would call the white people on campus out to a place called
White Plaza. The white people would say: "RIGHT ON." I
thought then, "well nothing good can ever come of this" because there are two sets of racial stereotypes involved. If you
assume that all the good people are on one side and all the bad
people are on the other side, you are just missing the point
about what it is to be human. That is, we are each a mixture
of good and bad, requiring some kind of judgment, sympathy,
and forgiveness. You know Stanford?

BH: I know it well. I went there as an undergraduate. I studied
writing there. I went to many a demonstration at White Plaza.

WB: Well, one day I was standing in the quadrangle and there was
some kind of demonstration taking place. I said to the friend
I was with, "What in the hell is going on?" Some white boy
who was going by picked up on my "southern speech," turned
around and said in perfect fury "Youdamn well better find
out!" I thought guilt and anger were the wrong motives for a
conversation about race. What I thought was missing was love.
And I started thinking about Nick and Aunt Georgie, a black
couple who were my friends when I was a boy. I couldn't speak
for them, you know, a white man can't stand and face the world
and say that these old black people loved him back in the 40s.
But I knew I loved them.

BH: The way that love guides you to stand for justice, to stand in
solidarity is one aspect you share in the memoir portion of *The
Hidden Wound* that I find especially moving. In these last five
years much of my work on race has focused on the transformative power of love as a force that can lead us to social change.
I am always moved by the particular moment in you book

where you are having the birthday party and Nick cannot come in the house because black and white can't socialize together. You choose to leave the house to be with him. That's just such a perfect metaphor for what it means to give up unearned white privilege.

WB: Well, it's complicated. I was using my manners and it was manners my white elders had taught me.

BH: Yet you used those manners to intervene on a situation that was charged with the hidden violence of racism. And whether you were aware of it or not, you showed what white people can do to challenge and change that violence. That's why *The Hidden Wound* is still a valuable book. You bring complexity to our understanding of the intimacy between white and black people, an intimacy that certainly was on one hand created by the circumstances of oppression and exploitation, but as you show in this work that did not preclude the possibility of deep and abiding connections of care happening between white and black people even within this structure of domination.

WB: Well, there was a conventional structure called segregation. We weren't calling it that. I don't think we were calling it anything. It was just for most people the way things were. Certain assumptions were made, certain judgments were made, that was kind of a background but what was actually happening was that more often than not people were dealing with each other as individuals and there were all kind of exceptions.

BH: By that you mean folk were not staying within the rigid boundaries racism had set for them. They were not staying in their place.

WB: Remarkable things were going on that were exceptions to the rule. Kindness was going back and forth. And meanness was going back and forth between individuals, actual people. And so it's not as if the abstract structure [of segregation] was a pattern you could lay the life of the time down on and pat it into a neat shape.

BH: That perspective of a humanizing relationship that often shadowed the constraints placed by racist domination may not have been shared by the black folks who were likely to be the targets of racial aggression if they did not "stay in their place." That perspective may be informed by white privilege. But certainly growing up in the world of intense racial apartheid and segregation I know white and black folks found ways to meet and form intimacy despite the insanity of racial domination. By intimacy, I mean the kind of acknowledgment and understanding that can be the basis for love. That's the connection you make in *The Hidden Wound*.

WB: In those days white people told funny stories about black people, and black people told funny stories about white people. Segregation was happening but it was happening to us as we were. I took this up again in the essay you quoted from earlier "The American Imagination of the Civil War." It's about the way local life has been blurred by stereotypes of all kinds going back to the Civil War era.

BH: I agree with you that we fail to render a holistic picture of what segregation was like if we only focus on oppressors and victims, but we must also be careful not to overstate the case, not to act as though the humanizing interactions white and black folks undermined the overall exploitative and oppressive structure. But like you, I believe we will understand race better if we look not just at the ways people were victimized but look also at the ways affirming ties of care, affection, and even love were developed within the context of segregation, that mutual emotional dependency you write about. That positive dependency breaks down when domination, the notion that whites are superior therefore deserving to rule over a servant class deemed inferior.

WB: That dependence was practical to some extent. We needed, or so we thought, those people to work for us. What isn't being acknowledged now is that white people are doing the same thing with Mexicans. As a racially designated subservient class, these people are hired to do the work we are "too good" to do. We don't want to do certain fundamental work for ourselves. This is debilitating for us. The situation now is worse because there is no intimacy between races. You hear people say "My Mexican or So-and-So's Mexicans."

BH: But it is as though they are talking about inanimate objects and not people, like you identify my house, my car. Unlike other ethnic groups, Asian, and most recently Mexicans who are the new serving class, there is little intimacy between these groups and the white worlds they have to serve or serve. White people aren't as fascinated by Mexicans in the same way that there has been this historical fascination with blackness. No matter how much people enjoy Mexican music, it's never going to have a profound impact on the American culture, its never going to create a cultural revolution the way African-American music, in all its forms has. That symbiotic bond between black and white America is still unique. But these day it seems to primarily produce heartbreak. Racism intensifies because negative stereotypes are the only way of knowing and relating to the "other" that most use. We saw this during the Katrina catastrophe. When the amazing culture black people, especially the poor, helped create in New Orleans is not acknowledged. And the poor are just represented in the media as helpless victims or defeated predators.

WB: Well, the catastrophe seemed to be an ideal platform for stereotypes rather than actual people.

BH: Whether we are talking about disenfranchised poor black folks or migrant workers, some of whom are still black, the point is the refusal on the part of dominator culture to acknowledge their humanity. And poor working Mexicans are now prime targets for this brutal dehumanization.

WB: The migrant laborers don't even have the protection of being [human] property. If they were property, maybe people would try to take care of them.

BH: To the extent that they are viewed as objects, they are disposable. In the segregated world of the American South, black folks could not be disposed of because they were seen as necessary for the making of life. And what we know is white folks do not see Mexican workers as integral to their life and culture. There is usually no emotional engagement there — no care. In dominator culture it is usually the folks who have the least who give care to those who have the most. Recently, my mother was placed in a nursing home for a short stay. Most of the residents were white. But there was a plantation culture happening there, with the serving class, black folks and the served whites. Almost all of the folk giving care, dressing and undressing residents, washing bodies, wiping bottoms, cleaning up mess are, serving, are black. And the folks giving orders whether as high-level administrators or as residents are white. Visiting Mama in the nursing home, I observed the racial hierarchy, whites at the top, black on the bottom, whites giving orders, black folks taking orders. And yet here again, this is the superficial picture; the reality is more complex. For here in this place of sickness and death, there is a profound dependency of white needing black, of white depending on the kindness of black strangers. Underneath the surface there is a culture where bonds are established, where folk talk across race. Still, in many places in our society a more inhuman plantation culture (where whites dominate black people and other groups) is the norm. It seems as though our nation has created a modern context for slavery. Do you think slavery has ended or has it simply taken new forms?

WB: Well, I think it has taken on new forms. A lot of white people are thinking of themselves as slaves, and some of them are "successful" people. You have a whole society that is saying, "Thank God. It's Friday." They are thinking of themselves as involuntary servants complicit in their own shackling.

BH: Yet they may think this way and have little concern or empathy for folks lower on the economic totem pole who really must work like slaves for inadequate wages. Migrant workers are the prime example.

WB: Why do we have these migrant workers? Because we [white people] think we are too good for physical work and physical reality.

BH: And many upwardly mobile black people feel the same; that's why they want not to be reminded of our agrarian past or of the plight of those black folks who are the working poor, who must make their living doing physical work.

WB: Some groups of white people, such as the Old Older Amish, take responsibility for themselves. I'm told that when an Amish person dies, the young women prepare the body and the young men dig the grave. Certain essential things are taught to the young people who do that work. I think we [non-Amish white folks] are a people who have always needed people to look down on, people to do what we used to call "nigger" work. Yet on the other hand, there was also a saying here among some folk: "I would never ask anyone to do anything I wouldn't do." Among farmers around here that saying was pointed directly at the racial structure. In other words — a part of my pride will be that I will not ask other people to lower themselves to do something I think I'm too good to do.

BH: For white folk who see certain kinds of work as beneath them, there has to always be a subordinated class to do the dirty work, whether that class has a racialized or ethnic identity. Sadly, many disenfranchised black and white poor people buy into this same logic and feel they are "too good" for certain forms of labor. Throughout my growing up the elders and my more modern parents were clear that "all honest work is good work." And we all taught that any labor done well, cleaning out a barn, the outhouse, cooking or serving done well, would be humanizing. This attitude towards work made black folk strong back in the day, a people of wisdom and integrity. Growing up the country black folk were a people of hard work, generosity, humility, and integrity. I see that wisdom in the portraits of Nick and Aunt Georgie in *The Hidden Wound*.

WB: That's right. Rural black people then were a people with wonderful knowledge, really essential knowledge. They knew how to make do, how to live on the margins. The chef Alice Waters has made a sort of revolution in our times by reminding people that good cooking depends on good ingredients.. On the farms around here, when Aunt Georgie was living, people would have been surprised to hear that there could be such a thing as bad ingredients — black and white; everybody had good ingredients. Cooking was uncommonly good everywhere you went. But the black people knew how to use the pieces of pork that the white people did not want to eat.

BH: I liken it to a culinary recycling. I can remember my grandmother laughing about the stuff white folks would throw away that they would take and make of it something mouthwatering, something delicious.

WB: Think of the beauty of their intimacy with the material life that they lived. They didn't have much, but everything they had they prized. Aunt Georgie was a great embroiderer and a great quiltmaker.

BH: Fortunately, I had a community of folks in my life like Nick and Aunt Geordie. I came in to the last of that holistic, organic world with my grandparents. I grew up in that world of farming, of sharecropping. Baba raised her chickens, made butt, made soap, and wine from our grapes. In my child's mind their world was a paradise. They worked hard. They lovedtheir land. And they shared that love. When I left our little town in Kentucky and went to Stanford and met all those black people there who thought they were too good to do basic work, I could not relate to them. Urban black culture, city culture was just beginning to be the yardstick against which everything about blackness would come to be defined. All the aspects of our identity and culture that was deemed relevant came from the city. Gone was a world where black folks understood the limitations of white power. My Daddy Jerry, my paternal grandfather, as he plowed with his mule would say; "You see that sun — the white man can't make it rise — no man can make it rise — man ain't everything." Daddy Jerry knew that there were limits to white power and to human power. We are living in a world right not where many black people and other people of color feel that white power is absolute. They see themselves as victims. They feel constant defeat and despair. In the culture of southern blackness, of Kentucky farm culture, you and I evoke, black folk were able to maintain integrity, dignity, creating beauty in the midst of exploitation and oppression. They did not give themselves over to sorrow. That did not mean they did not grieve. But even grief had its place. The important thing was to keep a hold on life.

WB: I think that this is what is most authentic about *The Hidden Wound*, the understanding that Nick and Georgie were admirable people.

BH: They were people of integrity. The root meaning of integrity is wholeness. And they are the people who are my teachers who are helping me by their example to be whole. Certainly loving these two black folks helped the little boy that you were to grown into a whole man. Look at what intimacy and genuine love can do.

WB: Well, strange things happen. I had an aunt who grew up down here, and she moved to Indianapolis. She and her next-door neighbor shared a black housemaid. My aunt's neighbor always set a place for the maid, and they sat and ate dinner together. When she came to my aunt's house, my aunt ate and then the black woman ate. One day the neighbor was at my aunt's house, and the maid got a call about a death in the family. She went all to pieces, as people do. My aunt just gathered her up in her arms and comforted her. The neighbor, who always ate with the maid asked, "How could you put your arms around that nigger." It's so mixed.

BH: This is why it has been difficult to honestly talk about race. We are surrounded by a profound silence about race. And the talk we hear, the public talk about race is usually just a pouring of gasoline onto the fire. Most if it does nothing to end racism. It's the profound silence that we live within because we lack a language that is complex enough. Our task as people who love justice is to create that language. And to affirm those social contexts where white and black folks bond beyond race.

WB: I know there are situations in which there is mutual respect and situations where there is no coming together. The churches are still segregated.

BH: Being in the church is also about being in the body. Religion so often determines which body will be seen as sacred, as worthy of life. It takes us right back to the earth you know, because in a sense that same kind of paradoxical relationship is what many Kentuckians have with the land. People, many of them Christians, can say they love Kentucky... its "unbridled spirit," but then be passive in the face of mountain-top removal, all the while talking about the beauty of these Kentucky hills and the majesty of our mountains. And folks outside Kentucky rarely understand. Many of them have never heard about mountain-top removal. And how many Americans even think about coal — where it comes from — what it is used for. That's why everything you write about industrial agriculture is so important. It's why *The Unsettling of America* is still a book that opens our eyes.

WB: In an interview with Rose Berger ...

BH: Oh, yes the one called "Heaven In Henry County." (shared laughter)

WB: ... I stated that: "It's sort of normal to wish that things like that would not be applicable any longer and it's discouraging to see that they stay current. You want that a book like *The Unsettling of American* would become obsolete, but it's more relevant now than it ever was. ... Look at the way we mine coal, for instance. Look at the way we're logging the forests. These are not sustainable procedures. They're not conservative procedures."

BH: In *The Hidden Wound* you talk about racism as a "disorder of the heart" and that's just a beautifully way to name this pathology. And that sense of disorder governs most of our citizens' relationship to land, to development. Just as the internalized racism of black people makes many of us terribly complicit in this system of domination, racism cannot shield those black people who follow the ecological madness of the mainstream. Certainly black folks have been complicit with the erasure of our agrarian past. George Washington Carver is still one of the most visionary conservationists. Devoting his life to issues of sustainability, he wanted to teach everyone but especially farmers, white and black, how to tend the soil, how to grow appropriate crops. He wanted to save and preserve the way of life of the small farmer, of any race.

WB: Booker T. Washington had his share of trouble attempting to teach folks to value the land. Remember the statement he made that people resented was: "Cast down your buckets where you are."

BH: Washington was different from Carver who was a practitioner. Carver was hands on. Every day of his life he went out into nature, working with the earth. His seemingly mystic communion with nature was a way to be guided by divine spirit. Washington was led by politics. When we read Carver he's talking about working the land but he is also talking about finding peace in communion with nature. He's talking about the spirit of land and what's going to happen when we learn to use the land in a way that is not destruction. Few schoolchildren today know about Carver, all that he did with the peanut and the sweet potato to create a revolution for the farmer in the deep South. Erasing this knowledge enhances white supremacy. Environmentalists, mostly white people, will say they can't get black folk to be concerned about ecology or sustainable. But this is another racial silence. How many black people were socialized to devalue working land — to look down on farming? Whose interest has it served to deny our agrarian African-American past? At what point did we stop hearing the language of the black farmer? I had that language growing up. What's happened to that language is systematic destruction! It has served the interest of imperialist white supremacist capitalist patriarchy to teach everyone, especially black people, that there is only one world that matters — the world of the city, of buying, of things. That's why I teach *The Hidden Wound* as often as I can because it documents that agrarian past. There is always a degree of tension in the class because there are always students who want to know why I think this book is relevant to today. I especially want students to read about the black elders you knew and loved as a boy, to read about their deep sense of integrity of being. That's something most Americans, including black folks, don't know about.

WB: People are ashamed of that part of our heritage and to refuse to own it is a mistake. If we ever fully recognize our past, we will understand how deeply we need each other. One of the things I remember and think about is how well the agrarian black folk

knew this county at night because they hunted then. No one
expected them to work at night, so then they were free. They
might have been hunting to put food on the table or just going
out to hear their dogs run a fox. Nick always has a fox hound
or two.

BH: A fox hound and a coon dog, Daddy Jerry loved to prowl the
night. When the poet writes: "I have been on acquainted with
the night." I think of those escaped slaves, then those black
male farmers who found the dark a place to roam, a place of
freedom. Daddy Jerry always tried to get his grandchildren
to come out in the pitch dark "to learn the dark" — to learn
its comforts and its solace. We can do that and learn to be
comfortable in the darkness and beauty of our skin. No one
can take that spirit of belonging away. That power in loving
blackness was there in my childhood, and I learned these lessons from black and white folks. The world of shared work
brought folks together across the boundaries of race. We hear
these true life stories from coal miners and their family who
say, "When we went down in that coal pit, it didn't matter if
you were black or white because we all black. We all come out
black." People who do not know that black men and women
worked in the mines need to go to Lynch, Kentucky. Again,
in this work there was a sense of the urgency of life and death.
That when facing death false constructions like race cease to
matter. Today we are living in a culture where that urgency is
not acknowledged. People believe they can control, that they
can escape. They really believe they can destroy nature and
succeed. They believe that they can survive and thrive. This is
the insanity we live with.

WB: When black men and white men mine coal together they are in
mutually useful relationships. They depend on each other. This
relationship is urgent at times. If there were white and black
people on the same farm, you saw the same thing. Someone,

whoever he was, would get the reputation of being a good hand wherever you put him. You didn't need to worry about him. He'd be there.

BH: And he would work long and hard. When I entered classrooms at Stanford University and I would hear stories about black men, their laziness and their corruption. I felt it was all surreal. I did not know too many black men like the ones they described. Every black man I had known in my life up until that point had been a worker on the earth, a hard worker. As a child I was fascinated by the hands of grandfathers. These were hands that plowed, planted, and prayed, hands that would protect me, that would embrace and caress when the days work was done. I would touch every crevice of these big calloused black hands. So, when I went to college and heard all these negative stereotypes about black males, I was stunned. Growing into the realization that to recover the experience of these glorious agrarian black male has deepened my work; it is my way to pay tribute. That's why I think Kentucky is my fate. It is my calling to remember their hard work and along with others tell their stories.

WB: There is ecstasy in that kind of work. Ernest Gaines understands this. He understands what that accomplishment meant. Another thing is that agrarian black males often had their own domestic animals. If you were a black hand around this part of the country and you were living on somebody's farm, working by the day, raising a crop, or whatever you did, you would always work with your own team. There is a complex relationship between a person and a team. It can be a transfiguring relationship when you are calling on them and your team does what you want them to do and do it beautifully. There is a great beauty involved in that. It's horrible that there are all these bodies that are not useful for anything. When I see people on treadmills I think, if you had that human power working on a farm, you might be able to clear out a fence row.

BH: Most people imagine that black folks working the land were just victims, working for little and living a starvation life. We both know that the life of a small farmer can be terribly hard. What outsiders rarely see is the spiritual reward — the power of redemptive suffering. When you live in a capitalist culture that tells you all forms of suffering is bad (take this pill, this shot, have this operation, make the pain go away), then you lose the mystery and magic of redemptive suffering. When her children were sold from her, Sojourner Truth could declare: "When I cried out with a mother's grief none but Jesus heard." She's crying out in the woods. She's down on her knees surrounded by a powerful natural world. And there in her moment of profound grief she finds solace. Hers is a mystical moment of union with the divine. Much of what our nation has lost is that awareness that the earth can be for us a place of spiritual renewal, not just a place to stroll in a park, or hike in a forest, or find land to mine resources, but that it is a place where we can be transformed. As black people go back to the testimony of Nick and Aunt Georgie, and black folks like them living and dead, we can remember and learn from them. Their strength came from knowing they could look at the hills and be restored — that no matter the deathly deeds of humans the earth would stand as their eternal witness.

19

Take Back the Night — Remake the Present

I have always come home to Kentucky but I was just visiting. Now I have come home to stay — to stay forever is what I dream about, even though I know that everyday dreams change. Coming back to my native place once or twice a year as I did for thirty years, to the eclectic strange world of small-town Kentucky, I was welcomed with open arms and those same arms held me close as I was leaving. All those years Rosa Bell and Veodis, Mama and Daddy, would begin the slow process of saying goodbye by walking with me out the door into the yard. There they would stand watching, holding onto their farewells as a musician holds onto a musical note until its reaches an end — until those leaving and those left behind can see each other no more. All the time leaving and returning, never staying in one place, I carry in my mind's eye this snapshot of my parents standing in one place, standing in their marriage of more than fifty years, standing in the familiar homeplace. In the world of our growing up it was deemed vital for one's well-being to stay in one place. Poet Gary Snyder says that there comes a time in life when you have to "stop somewhere." I have stopped here in small-town Kentucky. I am staying in place.

A true home is the place — any place — where growth is nurtured, where there is constancy. As much as change is always happening whether we want it to or not, there is still a need we have for constancy. Our first home is the earth, and it will be where we come again to rest forever, our final homeplace. The red clay dirt I ate as a child, being told "you gotta eat dirt before you die," would keep me familiar with the dying process. As though eating dirt helped make one ready to be at home in the grave. This red dirt that was the ground of my being and becoming was a color more typically found in the terrain of the southwest and other desert landscapes. Here in Kentucky, it was special, sacred, part of a magical landscape.

Giddy with delight as I walk in the wet mud in the back of my hill house, I feel as ecstatic on this Kentucky ground as I felt as a child innocently confident that the earth was a gigantic playground. Surrounded by trees (climbing trees was part of childhood fun) stumbling down the hill, I am reminded of that moment in childhood when our brother convinced me and my younger sister to lie down and curl our bodies so that like a ball he could roll us down the hill. My younger sister carries the scars from that adventure to this day. I sit at the top of a hill as I once did as a child and give thanks that this powerful experience of the healing power of nature is not just nostalgic and sentimental reflections on the past that in my present I experience this healing power once again.

When we were not roaming the hills in our childhood, we were walking and playing in meadows. An open ongoing meadow was my field of dreams as a child. In that expanse of land and magical growing things I felt sublimely blissful, able to sit for hours, to lie back and soak in the sky, to feel the coolness of earth on my back and the heat of the sun covering the front of me. This was heaven — this

world of weeds, wildflowers, wild berries, and asparagus. Later in
life, lines from a poem by Robert Duncan, "often I am permitted
to return to a meadow," evoke for memories of contentment. These
words conjure for me an active nostalgia in which imaginative memory enables one to return to the state of mind evoked by a place
even if you are no longer able to return to that place. Now in my
Kentucky present there is always a meadow to return to, a place to
sit, wonder, contemplate.

Folks who stay away from their native place, never visiting, who
then return, who come home (so to speak), inhabit a different psychic landscape from those of us who were constantly coming home,
constantly conflicted about whether we will go or stay. Fortunately,
I sustained ties with home and family, resisting the urge to break
bonds when family was not as I wanted it to be. I felt those ties bind-
ing and holding me no matter where I traveled in the world. Since
there was no violent separation between me and my past, my family, I felt no need to destroy all seeds of hope within me that might
nurture a will to return home. Leaving home and staying away felt
like betrayal, especially when my maternal grandmother who embodied many of the old ways would question me about how it was I
could live so far from my people. While I lived far from my family
in miles, they constantly inhabited the space of my dreaming. They
followed me everywhere, telling me how I should live. Ironically, I
had no life to live. The spirit ancestors must have known this. They
watched over me, guarding my way, as I struggled to find myself, to
find my way home.

Even though life in the dysfunctional primary family was an endless series of hurts and heartbreaks, all the knowledge and wisdom
that was shared with me by the old ones, the elders, from church
and community, was empowering and amazing. From my elders
I gained the strength of character to act with courage and integrity. From them I learned the importance of listening to one's spirit
guides, to see those guides as the way the ancestors speak to and

through us. While I left home fleeing the soul murder that had left me feeling abandoned and lacking, I begin to realize as I wandered from place to place, trying to find myself, that I had been given these precious gifts from the elders that enabled me to survive and thrive. Talking and writing, again and again, about this received wisdom is essential for those of us who not only want to remember the old ways, who want to integrate this wisdom from the past into our present as it remains life sustaining even though many of the elder teachers are long gone.

In my family of origin both my parents were negative about the old ways. They wanted a modern life, a life structured around the principles of liberal individualism, a life where the fulfillment of material desires mattered most. They refused to acknowledge the value of oppositional ways of thinking and being black folk had created in the segregated sub-cultures of Kentucky, especially the backwoods culture. More than anything our parents wanted their children to conform. I was intrigued by the culture of non-conformity, by the outlaw culture of my maternal grandparents. They believed that being a person of integrity was the most important aspect of anyone's life. Second, they believed that it was essential to be self-determining and self-reliant. Since they lived life guided by the principles of organic environmental sustainability planting flowers and growing their food, raising their animals, digging fishing worms, making soap, wine, quilts, they were never wasteful. The focus of their life was meeting basic needs, keeping the wisdom of living off the land. They believed in the value of land ownership because owning one's land was all that made self-determination possible. Although their lives were fraught with difficulties, especially as they daily encountered a world where their values had little meaning, the essence of all they were teaching and being holds true.

All the elders in my life growing up, whether they were family or chosen kin, believed it was essential for us to have a spiritual foundation. While Christianity was given pride of place in the quest for religious allegiance, my paternal grandmother Sister Ray believed in the power of voodoo, often ridiculed as hoodoo. No matter their choice for spiritual direction, our spiritually aware elders accepted that one could have mystical experiences based on communion with divine spirit. Both in slavery and beyond, individuals would often go on solitary retreats in their quest for divine guidance and intervention. In African-American history Sojourner Truth is one of the most well-known anti-slavery freedom fighters who gave testimony to her spiritual visions and mystical experience. During their communion with nature, country black folks who were not consciously seeking for inner mystical experience found themselves undergoing a shift in consciousness. That shift enabled them to feel oneness with all of creation giving them a sense of well-being, bliss, and a wise understanding of the reality of impermanence. This expanded sense of spiritual awareness moved far beyond the constraints of Christian thought and doctrine. It called on believers to acknowledge both psychic power, intuition, and the power of the unconscious.

When I was growing up we would hear grown-ups talk about the individual women who could predict the future and who could make things happen for you, individuals who were seers and healers. They could tell you the meaning of your dreams. Believing in the importance of dream and dream interpretation, many of our elders, like Sister Ray, acknowledged the power of mind, the subconscious. They believed that by listening to the messages given us in dreams one could be guided in daily life. They also believed in the importance of intuition. Guided by intuitions one could foresee a future reality and be proactive in relationship to it. All of these beliefs, acceptance of the oneness of life, the necessity of spiritual awareness, and the willingness to follow divine guidance, all helped sustain the belief in transcendence, in a cosmic consciousness more powerful than humankind. It kept black folks living in the midst of racial apartheid from being overwhelmed by despair. It kept them from seeing themselves as always and only victims. To believe in transcendence gave one a concrete basis for hope, for remembering that change is always possible. These empowering aspects of African-American southern life that were commonplace in segregated black communities began to lose their appeal as folks sought to integrate mainstream society and become part of dominator culture.

While my grandparents looked at most white folks with a critical eye, seeing them, more often than not, as pathologically narcissistic, racial integration ushered in a world wherein many black folks were trying to live as white folks as lived, to be like them. Creating standards of being and becoming that were not rooted in a quest to have what white folks had (of course when black folks expressed such longings they were not talking about having what poor white people had, they were equating whiteness with privilege) was part of oppositional thinking. It allowed for the formation of a different life principle which offered different ways of thinking and being. Ironically, despite the fact that our elders who lived in segregation suffered more intense race and class-based exploitation and oppression than most black folks experience today, they had a foundation for building better healthy self-esteem than those of us raised in the world of racial integration and greater economic opportunity. In many ways their lives were hard, and yet they had ways of knowing joy and peace that are not known to many black people living in present times.

Significantly, it cannot be stated enough that the sense of oneness with nature which offered a transcendental sense of life wherein humans were simply a small part of the holistic picture helped agrarian black folk put notions of race and racial superiority in perspective. In the segregated world of my growing up, black folks did not think that whiteness was all-powerful. We were raised to see exploitation and oppression of any group as a sign of moral depravity. White people were in general not admired or envied. While it was clear that they were usually better off materially or had greater access to economic mobility, they were not seen as having a better life.

Consequently, most black people did not see themselves as victims with no power to choose the quality of their life. Meeting adversity with perseverance and learning how to cope with difficulties shaped the content of one's character. And one's character determined one's fate.

Creating joy in the midst of adversity was an essential survival strategy. More often than not peace and happiness were found in the enjoyment of simplicity. The pleasure of ripe fruit, a good tomato, smoking tobacco that one had grown, cured, and rolled into cigarettes, hunting, or catching fish. These simple pleasures created the context for contentment. Calling to mind these earlier times in African-American life and culture is not a sentimental gesture or an expression of empty nostalgia, it is meant to remind those of us grappling with the construction of self and identity in the present that we have a legacy of positive survival skills to draw upon that can teach us how to live with optimal well-being, irrespective of our circumstance. Suppressing these insights, erasing the agrarian roots of African-American folk, was a strategy of domination and colonization used by imperialist white supremacist capitalists to make it impossible for black people to choose self-determination. Equating freedom solely with economic mobility and material acquisition was a way of thinking about life that led country black folk to seek to distance themselves from their agrarian past. Eventually, this process of forgetting the past was helped by the invention of sociology as an academic discipline which led to studies of black life that defined black identity solely in relation to urban experience. Fleeing their agrarian roots, most blacks left behind the oppositional values that had been a source of power, a culture of resistance based on alternative ways of living one that valued emotional intelligence.

Racial integration disrupted black sub-cultures by compelling conformity to mainstream dominant culture. This strategy of acculturation began with education. Prior to the integration of schools, there was a connection between the values black folks learned in our community and the values learned in schools. In the all-black schools of our growing up the wisdom of elderly black folks was acknowledged. Teachers would reference life lessons taught in the home. When schools were integrated, biased teachers encouraged us to believe that only white people had access to metaphysical knowledge and understanding. The organic metaphysics of our ancestors had no place in the "white schools." Often times smart black children who resisted socialization in the white schools were in a constant state of psychological trauma engendered by ongoing conflict with white power. On one hand, one had to "wear the mask" to succeed while trying to hold onto the unique culture of resistance formed in all-black settings.

Ironically, it was during my time away from home when I lived in predominantly white educational settings that I begin to rediscover and reclaim many of the alternative ways of thinking and being that I had learned growing up. By then, much of the oppositional culture I cherished was being erased. The culture of belonging was no longer common in black sub-cultures. This giving over was a direct consequence of efforts to succeed in dominator culture and one could be successful there only if one imitated the behavior of those in power. Oftentimes black people who could not conform saw no alternative and sought escape in addiction. The focus on white power overshadowed ancestral understanding that we were all more than "race" and that there were powers mightier than humans. The absence of this oppositional awareness led to a widespread feeling of vulnerability wherein many black people began to think of themselves solely as victims.

As a tool of brainwashing, television played a major role in the colonization of the black mind. It was television that brought the thinking of dominator culture into the homes of black folks. It imaged for us liberal individualism. No wonder then, as black folks embraced the thinking of dominator culture, that many of the more healthy ways of living that had been central to a culture of resistance (especially in the southern parts of our nation where racism was more overt and extreme) were no longer deemed valuable. And as black identity became more and more equated with urban experience, the more complex and multifaceted life experiences of southern black folk received little or no acknowledgment from mainstream culture. The portrait of southern black experience that emerged during the heady years of civil rights struggle was often one wherein a dehumanizing image of abject poverty was depicted. It was only as black women writers began to receive national attention in the wake of feminist movement that writers like Zora Neale Hurston, Alice Walker, Toni Morrison, and Gloria Naylor offered a more expansive vision of southern black life.

Hurston described segregated life in Florida in which artistic production flourished, especially the creation of music and dance, despite the harsh background of material lack and seemingly endless hard labor in oppressive circumstances. In her anthropological work Hurston highlighted southern black folk tales and showed the ways they were used to convey alternative ways of thinking and being to dominator culture. Walker wrote non-fiction essays about the beauty of her mother's gardens, about the growing of vegetables. Contrasting southern experience with northern experience in her first novel, *The Bluest Eye*, Morrison evokes a southern culture wherein black folk find their humanity in the natural environment, a culture that is sensual and life-enhancing. The life of southern black folk as she describes it is enhanced through luminosity. In the natural landscape everything is more vivid, radiant. Morrison's character Miss Pauline remembers her life in the South through colors: "When I first seed Chollly, I want you to know it was like all the bits of color from that time down home when all us chil'ren went berry picking after a funeral and I put some in the pocket of my Sunday dress, and they mashed up and stained my hips. My whole dress was messed with purple … I can feel that purple deep inside me…. And the streak of green them june bugs made on all the trees … All them colors was in, just sitting there." The relationship to nature she describes is one in which a connection is felt with everything, with the oneness of all life.

Walker's humanization of the oppressive black male Mister in her bestselling novel *The Color Purple* occurs as he experiences the wonder of nature, the beauty of trees and flowers. Reflecting on the meaning of life, he comments, "I think us here to wonder, myself. To wonder. To ast. And that in wondering bout the big things and asting bout the big thins, you learn about the little ones, almost by accident. ... The more I wonder, he say, the more I love." Returning to nature he is able to experience "wonder" and dream and by doing so open his heart. Gloria Naylor's *Mama Day* chronicles the return of the successful black professional woman coming back to her southern roots to find healing and happiness. Drawing on the tradition of "illumination," she fictively explores the way alternative spiritual traditions combine with traditional Christian ritual and are passed down from generation to generation. Black, male novelist Ernest Gaines received national attention when his novel, *A Gathering of Old Men*, was chosen as an Oprah Book Club selection. Fictively depicting the efforts of black folks to keep alive their agrarian legacy, their stewardship in relationship to the land, Gaines highlights the role of black farmers and the attempt by dominator culture to erase that history.

More recently art world focus on the quilts made by the all-black communities of Gees Bend, Alabama offers a glimpse of a culture of resistance that emerged in the midst of ongoing material deprivation and harsh circumstances, both public and private. Beautiful art books highlighting the work of the Gees Bend quilters use autobiographical narratives of individual black women to convey their way of life, the culture of giving and sharing that was common in their environment, the ability to live simply, to grow one's food, to work hard, and yet create beauty despite adversity. In all the narratives of the black women artists of Gees Bend, there is powerful testimony about the way in which spiritual awareness provided the basis for a new vision of life which first manifested in creative energy and expression. It is impossible to read their testimony and not be awed by the passion and skill these women brought and bring to artistic production even when there was or is little or no financial reward for the marvelous visions they created and create with scraps of cloth.

Significantly, when mainstream culture highlights the culture and traditions of southern black folk, it is usually to reveal its beauty, then to announce its passing. Rarely does anyone suggest that individual black folk today might benefit from reclaiming their roots to the land, that they might find migrating from the city to the country restorative, or find emotional sustenance in alternative spiritual practices, or just basic health by returning to a diet of home-grown and homemade foods. Instead, traditional southern black folk culture tends to be viewed solely from the standpoint of what can be seen as a brand of flat sentimental nostalgia, that is usually expressed in everyday reference to how things were much better in the past, "back in the day," but with no attempt to attempt to integrate all that was constructive and positive in the past with our lives in the present.

It has been especially difficult for black folks, whether living in the country or the city, to stop worshiping at the throne of secular economics, believing that just having access to greater material resources would or will make life grand. And even though we witness individual black people who have gained unprecedented enormous wealth yet testify to a lack of meaning and direction in their lives, the vast majority of black people (like almost everyone in our culture) still cling to the assumption that economic reward is the key to a good life. The celebration of folk roots and folk traditions seemed to capture the attention of affluent radical and/or new age white folks more than any other group. Even though there remains a focus on alternative living and new age spirituality that emerged in the late sixties and continues into the present day, it does not garner the interest of masses of our nation's citizens. It exists alongside and at times competes with a materialist worldview.

New age spirituality, alternative therapy, concern for diet and the environment are the sites wherein individuals are encouraged to go back to the land, to live a simple life. Even so, these cultural arenas are not heavily populated by people of color. While there are individual black folk who are returning from city to country, who are leaving the North to come South (these changes are documented by anthropological and sociological work on return migration), there are not many public forums where our passion for the environment, for local food production, spiritual awareness and living simply can gain a hearing. Integrating these concerns with traditional ways of knowing received from enlightened elders is important to some of us and we are trying to go back to that place of wisdom. We know that many important values of our past are in danger of being utterly lost if we do recover and reclaim them.

One of the great tools of colonization has been pushing the assumption that poor people (especially black people) have neither the inclination nor the time to be concerned about the substantive quality of their lives. Of course, this is one of the assumptions that would prove to be totally erroneous if there were more available information about our agrarian history. Mass media is certainly one forum we can use to teach and remind people. Films like Julie Dash's *Daughters of the Dust* and John Sayles' *American Beach* call attention to black engagement with land, with environmental concerns, with the global issue of sustainability.

Hopefully, as more black thinkers, writers, and artists share our engagement with the issues of environmental protection, local food production (both as consumers and producers), land stewardship, living simply, and our varied spiritual practices, we can chart a path that others will follow. Returning to one's native place is not an option for everyone but that does not mean that meaningful traditions and values that may have been a part of their past cannot be integrated into homeplace wherever they make it. In Bill Holm's book, *The Heart Can Be Filled Anywhere On Earth*, he recounts his own effort to return home after years of being away. For him it was the recognition, after leaving many places both home and abroad, that he had a longing "for a sense of from-ness, whatever its discomfort," that underlying that longing "was the desire for connection." He ends the introduction to these essays with this statement of intent about the their purpose: "they argue first that we are sunk by greed — consumerism gone mad, a mania to acquire what we neither need nor desire; by fear — of the 'stranger' who is only a disguise for fear of ourselves and our own history; by technology — which since we misuse it by trusting it too much, deracinates and abstracts us, separates us both from nature and each other; and finally by the mad notion that we define and invent ourselves in isolation from any sense of from-ness or connection." Whether reading Bill Holm or the more recent work of Barbara Kingsolver, *Animal, Vegetable, Miracle,* for many of us returning to our native place does bring an end to isolation. We are connected. And those connections both past and present solace, keep us excited by mystery, bring us joy.

20

Habits of the Heart

All my life I have searched for a place of belonging, a place that would become home. Growing up in a small Kentucky town, I knew in early childhood what home was, what it felt like. Home was the safe place, the place where one could count on not being hurt. It was the place where wounds were attended to. Home was the place where the me of me mattered. Home was the place I longed for; it was not where I lived. My first remembered family dwelling, a cinder block house with concrete floors on a hill, stood as though naked against the lush backdrop of a dense natural landscape: trees, honeysuckle vines, blackberry bushes, and wild strawberries all made the concrete house seem out of place, set against nature, but unable to take over the world of lush wild things since the house was fixed unchanging and the natural landscape adamantly growing.

In this wilderness where I first moved and lived and had my being, I was nature and nature was me. Nature was the intimate companion of my girlhood. When life inside the concrete house was painful, unbearable, there was always the outside. There was always a place for me in nature.

Over and over again the grown-ups would tell us to respect the wilderness around us, to understand that it could be friend or foe. Our task was to learn discernment, to be in nature as nature, to understand the limits of the natural world and of the human body in that world. Nature's generosity made it possible for us to have the pleasure of walking through fields and fields of homegrown vegetables, the pleasure of popping fresh yellow and red tomatoes into one's mouth straight from the vine. In this early childhood I experienced firsthand all that poet Gerard Manley Hopkins evokes when he writes that "nature is never spent," that within it "lives the dearest freshest deep down things." As a young child I believed the wilderness land around me had its own special perfume, that when I stayed outside for a long while that scent entered me and came with me indoors, the scent of a fecund world of growth reckless and without boundaries.

Roaming those Kentucky hills I dwelled in paradise. I was certain that I belonged confident of my place and purpose. In that culture of belonging, I learned the importance of divine providence. Walking behind Daddy Jerry, my father's father, farmer and sharecropper, I was taught that man could only do so much, that man could not make the crops grow, or the rain fall. I learned that humankind was special in our differences from other animals but that we were governed by higher powers. Daddy Jerry would often say, "as long as man knows his place in nature, everything will be right, but when he forgets and thinks he is god, trouble comes."

This profound belief in divine order allowed Daddy Jerry to experience wholeness and integrity despite the forces of white supremacist exploitation and oppression surrounding him. His love of the soil, the solace that he found in nature enabled him to have an open mind and an uplifted heart. Despite the sufferings he experienced living in the world of Jim Crow, subjected to the cruel whims of a white supremacist patriarchal regime, he found a culture of belonging in the natural world, with the earth as his witness. It was this culture of belonging he shared with me, his first granddaughter trailing behind him as he dropped seeds into the earth, as he harvested the fruits of his labor. In *Rebalancing the World* Carol Lee Flinders shares the insight that it is useful to think of the values of belonging as habits of the heart. It is fitting that in those early years of childhood I felt heartwhole.

Explaining further her understanding of a culture of belonging, Flinders writes: "The values of Belonging are, in effect, the symptoms of a particular way of being in the world. Together, they form a dynamic whole — a syndrome if you will, or an orientation or ethos. Within that whole, each value reinforces and all but implies the others, and the source of their power as a constellation is the synergy between them." Flinders asks readers to think of the values of belonging as "points on a circle, windows onto a single reality." Listing the characteristics of the culture of belonging, Flinders explains that: "Fundamental to the culture of belonging is a strong sense of reciprocal connection to the land where one lives, empathic relationship to animals, self restraint, custodial conservatism, deliberateness, balance, expressiveness, generosity, egalitarianism, mutuality affinity for alternative modes of knowing, playfulness, inclusiveness, nonviolent conflict resolution and openness to spirit."

The values of belonging imprinted on my consciousness in early childhood as a child of nature were in conflict with the values and beliefs that prevailed inside our patriarchal domestic household. In the concrete house I did not belong, there my spirit was alien, there I was subjected to soul-murdering assault. When our family moved from the hills, from the country, into town, it was a shift brought about by our mother's desire to have us be more civilized, to rid us of the taint of being from the backwoods, from being country people. Coming from a backwoods family who worked the land, growing their organic food, canning, raising chickens, making soap and wine, Mama wanted to get as far away from country ways as possible. That this move from country to city shattered my inner peace was all the evidence she needed to prove her argument that living in the hills was making her children strange.

For me, this move was traumatic. Trapped in my grief about leaving the natural landscape of my formative years, I became quite dysfunctional in the town, my sadness steady and constant. In the world of the town I was faced with the politics of race, class, and gender. From roaming hills and feeling free I learned in the world of the city that to be safe as a girl, and especially as a black girl, it was best to be still, enclosed, confined. I learned that to be safe within the space of blackness one had to keep within set boundaries, to not cross the tracks separating black from white. I learned that wearing homemade clothes and hand-me-downs were marks of shame. Gone was my confidence that I belonged in the world. Gone was the spirit of wildness rising in my soul each day like wind, like breath, like being. Explaining the significance of wildness in his collection of essays, *Hunting For Hope*, Scott Russell Sanders contends: "Like the trickster figure who show up in tales the world over, wildness has many guises, but chief among them are creator and destroyer ... Every form that gathers into existence eventually dissolves, every cell, every star ... Each heart that beats will one day cease. Knowing this, we have the choice of judging wildness, the very condition of our being, primarily by what it snatches away, of by what it gives." It was the generosity of wildness, receiving me, allowing me to be whole that led me to lament its loss in my young life.

In a world where I did not belong, I struggled to find strategies for survival. In the world of dominator culture, both within our household and beyond, I found a place of refuge in books, ways of perceiving the world which expanded my consciousness and left me wanting more from life than I believed was possible in the changing landscape of Kentucky as black people left hills, backwoods, and countryside for the promise of a better life in towns or left all together to migrate to northern cities.

Even though civil rights struggle integrated our high school, reunions were always segregated. When the time came for our twentieth reunion, it was decided that the races would be not be segregated, that the time had come for us to remember together. I sat at the table with the courageous white friends who had dared to cross the boundaries of race and class to make community. They shared with me their assumption that they always knew I would leave and not come back, that my soul was too large for the world of our growing up, for Kentucky. California and New York seemed to them the places that were right for me, the places that allowed one to be different and free. Like beloved black friends, they accepted that I would come home now and then but never to stay. Ann, the white female friend I wrote about in both *Bone Black* and *Wounds of Passion*, still lives in our town. Ken, our white male buddy, has a home not far away in which he lives some of the time and sees as the place to which he and his family will come to retire. In their adult lives they no longer have the intimacy with black friends that was there in our growing-up years. That they lead more segregated lives does not wound their spirits as it did when we were young and longed to live in community.

Oftentimes black folks who have left southern roots have symbolically run away to escape the everyday racism that restricts, limits, and confines, a racism that seems somehow worse than the racism one encounters elsewhere because it is intimate terrorism imposed not by strangers but by those with whom one is most familiar.

When I left Kentucky more than thirty years ago, I felt like an exile, as though I was being forced to leave the landscape of my origin, my native place because it would not allow me to grow, to be fully self-actualized. Both the inner world of family dysfunction and the outer world of dominator threatened to suffocate my spirit. Writing about exile in *Pedagogy of the Heart*, Paulo Freire contends: "Suffering exile implies recognizing that one has left his or her context of origin; it means experiencing bitterness, the clarity of a cloudy place where one must make right moves to get through. Exile cannot be suffered when it is all pain and pessimism. Exile cannot be suffered when it is all reason. One suffers exile when his or her conscious body, reason, and feeling — one's whole body — is touched ... To have a project for the future, I do not live only in the past. Rather, I exist in the present, where I prepare myself for the possible return." Exile highlighted for me all that was vital and life sustaining about the Kentucky world I grew up in. It was the field of dreams I explored to uncover the counterhegemonic culture of belonging that had made me different, able to be radically open. It was on Kentucky ground that I first experienced the interplay of race, gender, and class. It was there that I learned the importance of interlocking systems of domination. That experiential learning became a vital resource when I began to write critical theory.

Even though Kentucky and our vernacular tongue was the language of my dreams, I did not imagine myself returning to live in Kentucky. Time and time again I visited and felt that so little had changed. As a grown woman I felt that there was even less room for me here than when I was a child. Unchanged racism seemed to have grown more deeply entrenched. Plantation culture seemed to still be the norm. Narrow irrational fundamentalist religious thought continued to prevail. Neat justifications for domination culture were the order of the day.

To me, coming home was often like going back in time. And so many of the times that I returned were moments given over to grief that the positive old ways, the old culture, the old folk were leaving us. Throughout my work, both in visual art and writing, I have drawn on my childhood memories of life in Kentucky to evoke awareness of the power of a culture of belonging. To fully belong anywhere one must understand the ground of one's being. And that understanding invariably returns one to childhood. In *The Hidden Wound*, Kentucky writer Wendell Berry states: "I have gone back to my native place, to live there mindful of its nature and its possibilities." Even though Berry has lived in Kentucky most of his life, his reflections on the past are part of the effort to heal and become whole that is essential to the project of self-reclamation. Living away from my native place exploring the past and writing about it critically was a constant ritual of reclamation. It was a ritual, a remembering that not only evoked the past but made it a central part of the present. It was though I had not really left Kentucky as it was always there in imagination — the place I returned to — the ground of me being.

Researching return migration, the movement of African-Americans from urban cities to the rural South, anthropologist Carol Stack explains that the folks she interviewed wanted to come home to reclaim aspects of "belonging' and community they had not found in other places even as they also longed to work for change in the worlds of their growing up they seem so unchanged. She explains further: "No one is seeking timeless paradise; and no one, however nostalgic, is really seeking to turn back the clock ... What people are seeking is not so much the home they left behind as a place that they feel they can change, a place in which their lives and strivings will make a difference — a place in which to create a home." Obsessed with the project of creating home, I moved many places before making a decision to return to the South. Ultimately, I wanted to return to the place where I had felt myself to be part of a culture of belonging — to a place where I could feel at home, a landscape of memory, thought, and imagination.

For the more than thirty years I lived away from Kentucky; coming home to see my parents was always the ritual of regard renewing my sense of connection with the world of my growing up. My parents are old now. I can no longer count on them to be the force calling me home. Now the force within me demands that I stake my own claim to this Kentucky ground. The landscape of remembered belonging calls me to commune with the world of my growing up, the natural wildness that remains. Communing with nature is an essential aspect of a culture of belonging. In his book *Callings: Finding and Following An Authentic life*, Gregg Levoy reminds us that: "Nature is a proper setting for a return to ourselves, our source, our place of origin. It is the place where the world was created, where our ancestors came from ..."

Seeking solitude my spirit finds solace in nature. There it can embrace the reality of things living and dying, of the passing away of the old, of resurrection. Contemplation of dying and the deaths to come is yet another path that led me to come home again. The seventeenth-century Quaker William Penn counseled: "And this is the comfort of the good that the grave cannot hold them, and that they live as soon as they die. For death is no more than the turning us over from time to eternity. Death, then being the way and condition of life. We cannot love to live if we cannot bear to die." And so it is the knowledge of my own dying process that allows me to choose to return to a place where I first lived fully and well.

Certainly confronting death, the experience of 9/11, strangers and loved ones dying from disease and disaster, young and old, as well as facing the limitations of my aging body all awakened in me an intense yearning to experience anew cultures of belonging, even if they exist only as fragments, or just fledgling worlds trying to stay vibrant in the midst of dominator culture.

My return to my native place led me to Berea, a small town in eastern Kentucky with a progressive history and legacy. Berea College was founded in 1858 by a visionary abolitionist who believed in freedom for everyone, women and men. It was named after a biblical town in the New Testament "where people received the word with all readiness of mind." Fee founded Berea College with the expressed purpose of educating men and women, black and white from the Appalachian regions of Kentucky, especially poor folk. He wanted them to be able to learn in an environment embodying the principles of freedom, justice, and equality. He was committed to the creation of a lasting culture of belonging. It is fitting that I choose to create home in this place, to be part of a community working to sustain a culture of belong. Berea offers much that was wondrous in my life as a child.

The world of my childhood was a world of contrasts; on one hand a lush green landscape of fast horses, natural waterfalls, tobacco crops, and red birds and, on the other hand, a world of greedy exploitation of big homes and little shacks, a world of fear and domination, of man over nature, of white over black, or top and bottom. In my childhood I dreamed about a culture of belonging. I still dream that dream. I contemplate what our lives would be like if we knew how to cultivate awareness, to live mindfully, peacefully; if we learned habits of being that would bring us closer together, that would help us build beloved community. In my work rooted in my native place Kentucky and the values I learned as a child, I seek to evoke a language of healing, of hope, of possibility, a language of dreams, a language of belonging.

Awakening in the night, when I first moved to my new Kentucky home, I was startled by a familiar sound, the sound of a train, a sound evocative of my childhood. When we moved from country to city we lived only a few houses away from the train tracks. Every night I would lie in the stillness of the dark and listen to trains coming and going, imagining my own life journey, the places I would go, the people I would meet. The sound of the train comforts me now as it did then, for I know I have come home. I have returned to the world of my childhood, the world in which I first sowed the seeds of my being and becoming, a seeker on the path, the contemplative intellectual choosing solitude, ideas, choosing critical thinking. Here in my native place I embrace the circularity of the sacred, that where I begin is also where I will end. I belong here.

21

A Community of Care

Writing about my Kentucky past, I often say little about Rosa Bell (my mother) and Veodis (my father), yet their presence in Kentucky also called me home. Simply put, they were and are getting older, moving closer to death, and I wanted to spend time with them during their process of descent. My father has likened the period of life when one begins to be old as the time when we are no longer walking up the mountain. "Glory," he will say to me, "I'm never going to be walking up the mountain again, I'm going down the mountain. I'm on my way home." His metaphor astounds me because both Rosa and Veodis wanted to turn away from mountains and hills, to turn away from the agrarian life they had been born into and to seek after the modern and the new. No farming for them, no back-breaking labor on the land. They both wanted life in the city. And, as a child of the country, I have been at odds with them since my birth. Mama, sometimes jokingly and sometimes with rage, would rail against our many differences by exclaiming, "I don't know where I got you from but I sure wish I could take you back!" And oh how much I longed to go back, to go live with my grandparents with whom I felt a greater resonance of spirit. Mama and Daddy would not allow this.

They wanted me to become a city girl, and of all my siblings; they wanted me to be the one who would not be "country." And yet in many ways I am as country as they come, more like my grandparents than them. I even speak the language of my grandparents, the language of Kentucky black vernacular dialect but I also speak the language of the city, a language that is neutral with no attention to region or place. Hearing me speak the language of city was a comfort to my parents. That is until I acquired a dissident voice, one that shocked and jarred their sensibilities, a voice that made them feel afraid. To them any speaking out against authority, what I would call dominator culture, puts one at risk. And therefore it is better to remain silent. My talking made them afraid. In some ways they were glad then when I left home and went out into a world of cities where they did not have to hear me talk. They could never grasp that I was just plain country in lots of ways and that no amount of book learning, education, or writerly fame was going to change that.

In *Citizenship Papers*, Wendell Berry boldly states: "I believe that this contest between industrialism and agrarianism now defines the most fundamental human difference, for it divides not just two nearly opposite concepts of agriculture and land use, but also two nearly opposite ways of understanding ourselves, our fellow creatures, and our world." For me, this quote deeply evokes the schism between me and my parents. They represented the city, the culture of the new, "make more money, buy more things, throw things away, there is always more." My grandparents, both maternal and paternal represented the country, the culture of the old, no waste, everything used, useful, recycled. Now Rosa Bell and Veodis have themselves become part of the culture of the old. Dad at eighty-eight is one of the last living survivors of the all-black infantry he was part of in the Second World War. Mama is ten years younger but the loss of her memory has taken her from here to eternity. She, more so than Dad, feels that she has no real place among the living, that she does not belong. Unlike Dad, she feels it would be better to die.

Losing one's memory, to dementia or Alzheimer's, is a way of dying. It takes one to a place where you no longer make connections and communicate with the mind. Words no longer carry much weight. Language has little meaning. The divide between the country and the city no longer exists. Time cannot be understood in any consistent linear way. Time converges on itself; days past fall easily into the present, and years collapse upon themselves. Faces too fall into forgetfulness and relationships that were once all in all are indistinct shadows. Mama awakens and says of the husband she has been in partnership with for almost sixty years: "Who is he?" When I identify him, she just says, "Oh!" And that is where it ends for her. Later, she will call him by name and speak from that place where they know one another intimately. But this vivid awareness will not last.

Mama still knows who I am. She hears my voice and knows Gloria Jean is calling. She hears my voice and knows how I am feeling. One day I call and she says: "I was just looking at one of your books." When I was home last, she had one of my books and kept reading the part on the back that describes the author. Repeatedly, she reads it aloud to me over and over again. When she finishes, she is satisfied to have grasped a part of who I am, her daughter who writes. And yet my writing has been a source of pain to Mama, revealing to the public world much that she would have chosen to keep private, to keep secret. Even though she told me once that my work causes her so much pain that she just has to fall on her knees and pray, she is proud of my writing. Both my parents have weathered the storm of my work. Dysfunctional though our family may be, they have maintained their care and commitment to all their children — to family. And as I have grown into mid-life, I have come to appreciate deeply the discipline it takes to maintain commitment for more than fifty years. Living alone as I have done for almost as many years of my life as I shared with a partner, seeing marriages, partnerships (straight and gay) come and go, falling apart at the slightest moment where difference is recognized and deemed irreconcilable, I appreciate the strength it takes to maintain commitment.

I appreciate and understand a vision of marriage as a sacrament. P. Travis Krocker gives beautiful expression to what this means when he shares from a Christian standpoint that "giving ourselves away in marriage is an occasion of joy — we celebrate it because as humans we are made for intimate communion with God and with all of life." Explaining further he contends: "The sacrament of marriage is therefore anything but a private, exclusive act. It is always related to the larger community of which it is a part. One of the greater dangers of romantic love is that it privatizes love, depriving it of essential nutrients. A flourishing marriage needs the community to sustain it and will in turn build up the community and the life of the world." Indeed, I see this made evident in my parents' long marriage and saw it in the marriage of more than seventy years of my maternal grandparents. Sadly, both these marriages were not particularly loving or joyful. Even so the conditions for love were present: care, commitment, knowledge, responsibility, respect, and trust; all parties involved simply chose not to honor them in their wholeness. They chose instead to focus on care and commitment. As a witness to their lives, I can testify that they were fine disciplined examples of these two aspects of love. And despite the lack of sustained well-being in their marriages, I am still awed and impressed by the power of their will to commit. I long for such lifelong commitment in the context of a loving relationship.

Significantly, these two marriages lasted so long precisely because they took place in the context of community. They were buttressed by the constant interplay of life within extended family, church, work, and a civic world — a life in community. When I began to move past my harsh critiques of my parents' marriage, of their dysfunction, I could see positive aspects of their bonding. I could even feel envy. What I most admired and admire about their life is their capacity for disciplined commitment, their engagement in making and sustaining a life in community. And even though they did not create for themselves a loving bond, they did prepare the ground for love by sowing two important seeds, care and commitment, which I take to be essential to any effort to create love. Consequently, I am grateful to them for providing me and all their children an understanding via their life practice of what sustained commitment and care look like.

I am especially blessed to have lived long enough and to have parents still alive to whom I can express gratitude for their gifts of care and commitment. In his insightful essay "An Economy of Gratitude," Norman Wirzba shares this understanding: "In the practical, mundane, sustained commitment to place and community the marks of gratitude ... come into clearer focus." He defines those marks as "affection, attention, delight, kindness, praise, conviviality, and repentance." All these defining marks are present when I commune with our parents in our native place, in their Kentucky home. Gone is the spirit of conflict and contestation that for years characterized our interactions. By letting us know that there was no conflict that would be powerful enough to shatter ties of care and commitment, our parents, especially our mother, made it possible for there to always be a place of reconciliation, a place to come together, a way to return home.

Krocker emphasizes the importance of creating a "community of care" so that our relationships with one another can be "governed by conviviality rather than suspicion, by praise rather than blame." Furthermore: "In a community of care people are turned toward one another. They have given up the false, perpetually deferred dream that happiness lies somewhere else with other people." And that includes embracing our parents, accepting them for being who they are and not for becoming what we wanted or want them to be. Again Krocker explains: "As we work with others, and as we endeavor to get to know them, we learn to appreciate them in their depth and integrity and with a better appreciation for their potential and need. We see them for the unique creatures they are and begin to approach the complexity, beauty and mystery of every created thing and person. The loveliness of who they are starts to dawn on us, calling forth within us a response of love and celebration." Certainly, this has been my experience both in my relationship to my parents, the community of my growing up, and now to the place in Kentucky that is my home.

Communities of care are sustained by rituals of regard. Eating together was a central focus of family gatherings in our household. At the table we share our accounts of daily life, humor, and the sheer pleasure of delicious home-cooked food. Our mother was a great cook. I share with Krocker the belief that: "Around the table we create the conditions for conviviality and praise. In the sharing of the meal we give concrete expression to our gratitude. We catch a taste of heaven." This was certainly true in the kitchen of our mother's house.

Sadly, in her new state of lost memory, Mama no longer cooks, or finds delight in delicious eating. She has to be coaxed to come to the table. This is often the case with those who suffer dementia or Alzheimer's. New rituals of regard are needed. Before her memory loss, Mama was always on her feet working, cooking, cleaning, meeting someone else's needs. In their patriarchal marriage, she waited on our father hand and foot. Now she needs us to serve her, to dedicate ourselves to her comfort and care. This service is the enactment of a ritual of regard. The devotion she arouses in her loved ones is a natural outcome of the care and commitment she has extended to all of us. And even though it has been hard for Dad to change, to accept the end of certain forms of patriarchal privilege that he has assumed were his birthright simply by having been born male, he is learning to be a caregiver to her.

Nowadays, Mama spends much of her time sitting. There are beautiful and wondrous aspects to her current forms of self-expression and identity. It is a joy to sit next to her, to be able to hold her close, to caress her hands, all gestures that would have been impossible in the past. She would have deemed it silly to be sitting around talking of love when there was work to be done. How wonderful it is to have these new experiences with her converge with the old, to see her so tender, so vulnerable, so without the restraints of shame and conventional inhibitions. Now I see in her the wildness of spirit she once saw in me and wanted to crush for fear it was dangerous. My gratitude that I can be present — a witness to her life now — as she struggles to make sense of the dots that do not connect, as she journeys towards death, knows no bounds. Also it is good to witness Dad gracefully walking down the mountain giving him now and then a helping hand.

Krocker believes that "as we dedicate ourselves to one another, and thus experience daily and directly the diverse array of gifts that contribute to our living, gratitude will take its rightful place as the fundamental disposition that guides and forms our ways." Gratitude allows us to receive blessings; it prepares the ground of our being for love. And it is good to see that in the end, when all is said and done — love prevails.